Engineering Materials

For further volumes:
http://www.springer.com/series/4288

Jin Kuk Kim · Kaushik Pal

# Recent Advances in the Processing of Wood–Plastic Composites

Springer

Prof. Dr. Jin Kuk Kim
Department of Polymer Science and Engineering
School of Nano and Advanced Materials Engineering
Gyeongsang National University, Gazwa-Dong 900
Jinju, Gyeongnam, 600-701
South Korea
e-mail: rubber@gnu.ac.kr

Dr. Kaushik Pal
Department of Polymer Science and Engineering
School of Nano and Advanced Materials Engineering
Gyeongsang National University, Gazwa-Dong 900
Jinju, Gyeongnam, 600-701
South Korea
e-mail: pl_kshk@yahoo.co.in; pl.kshk@gmail.com

ISSN 1612-1317 e-ISSN 1868-1212

ISBN 978-3-642-14876-7 e-ISBN 978-3-642-14877-4

DOI 10.1007/978-3-642-14877-4

Springer Heidelberg Dordrecht London New York

Library of Congress Control Number: 2010934782



*Cover design:* deblik, Berlin

Printed on acid-free paper

Springer is part of Springer Science+Business Media (www.springer.com)

*For my parents, wife and childrens.*

**Jin Kuk Kim**

*For my parents, loving wife and teachers.*

**Kaushik Pal**

# Preface

Wood–plastic composite (commonly abbreviated as WPC) is a non-recyclable composite material lumber or timber made of recycled plastic and wood wastes and it has become one of the most dynamic sectors of the plastics industry in this decade. It is using in numerous applications, such as, outdoor deck floors, but it is also used for railings, fences, landscaping timbers, cladding and siding, park benches, molding and trim, window and door frames, and indoor furniture, etc. Scientists claim that wood–plastic composite is more environmentally friendly and requires less maintenance than the alternatives of solid wood treated with preservatives or solid wood of rot-resistant species. Resistant to cracking and splitting, these materials can be molded with or without simulated wood grain details. Even with the wood grain design these materials are still visually easy to distinguish from natural timber as the grains are the same uniform color as the rest of the material. Also, it can be recycled easily in a new wood–plastic composite.

Wood–plastic composite lumber is composed of wood from recovered saw dust (and other cellulose-based fiber fillers such as pulp fibers, peanut hulls, bamboo, straw, digestate, etc.) and virgin or waste plastics including high-density polyethylene, LDPE, HDPE, PVC, PP, ABS, PS, PLA, etc. The powder or fibers are mixed to a dough-like consistency and then extruded or molded to the desired shape. Additives such as colorants, coupling agents, stabilizers, blowing agents, reinforcing agents, foaming agents, lubricants help tailor the end product to the target area of application. Heat or UV stabilizers are added to prevent polymers from degradation from UV induced factors, while coupling agents improve wood resin adhesion. The material is formed into both solid and hollow profiles or into injection molded parts and products. With the diversity of organic components used in wood/plastic composite processing, there is no single answer to reliably handling these potentially difficult materials. In some applications standard thermoplastic injection moulding machines and tools can be utilized. Wood, resin, regrind, and most of the additives are combined and processed in a pelletizing extruder. The new material pellets are formed in mold and dried. Pre-distribution testing can help determine the optimal combination of chemical agents, design, agitation and other flow aid strategies for the specific material in use. Modern testing facilities are available to evaluate materials and determine the optimal combination of equipment components to assure the highest level of accuracy and reliability. Computerized performance test reports document equipment performance. Major advantage over wood is the ability of the material to be molded to meet almost any desired spatial conditions.

For convenience, the book is divided into four parts. The first, Chap. 1, relates to wood–plastic, and gives the reader a brief glimpse at the basic structures and properties of wood–plastic composites. Readers who are already familiar with wood–plastic composites through course work, research or engineering will find this material to be an elementary review. Although, detail has been kept to a minimum, we have attempted to describe the concepts of wood–plastic composites that are vital to an understanding of later sections.

The second part, Chap. 2 and 3, is concerned with surface treatment, machinery used for wood–plastic composites. The most exciting developments in the recent past about surface treatment and machinery used, is emphasized. A brief summary about the testing of wood–plastic composites materials is also summarized. Chapter 4 depicts about some recent work done in the recent past.

The third part, Chaps. 5–7, give a brief view of very recent developments, together with some unsolved problems. Broadly speaking these chapters deals with our lab work regarding the foam technology, flame retardant properties and colour retardant properties of the wood–plastic composites. The developments of most chapter proceeds from synthesis to morphology, and then shows how morphology affects or controls the physical and mechanical behaviours of the finished materials. It is interesting to note that significant proportions of each of the chapters are consumed in the form of wood–plastic composites blends, composites, or contributions of both. Thus, the reader will find references to apparently diverse industries side by side on the same page.

The fourth part, Chap. 8, gives an overview about some advanced applications of wood–plastic composites in our daily life. Mainly this chapter is a review work based on the daily use of wood–plastic composites in the present decade.

During the course of writing this book, it occurred to the author that a systematic classification of wood–plastic composite was sorely needed. In how many significantly different ways can wood–plastic be mixed with other polymers, or with non polymers? Also, it has been kept in attention about the work of WPC in foaming application. In this book the main aim is about the foaming technology used for WPC work. What interrelationship exists among the known modes, and how may we go about uncovering yet undiscovered combinations? While the classification theme pervades the text, weaving in and out, the actual classifications are left to Chaps. 5–8.

Every book is beamed at particular audience, we hope that chemists, chemical and mechanical engineers, polymer and materials scientists working in, or newly entering (or soon to be), the field of wood–plastic composites will constitute the prime audience for learning the basic principles by which one can analyze a wide variety of wood–plastic composite materials. Although, the book was not intended as a comprehensive book, it may well serve as a source book for the scientist wishing to work in this field. Readers who now work in the wood–plastic composite industries will surely find of their favourite process missing.

We wish to thank many people for their helpful discussions and sharing their expertise, Dr. V. Sreedhar of Gyeongsang National University for his useful contributions for this book. We appreciate Dr. Zhen Xiu Zhang for his Ph.D. work towards the blend of wood–plastic composites. We acknowledge to all the lab members in our lab for their kind contributions and generous support regarding the preparation and correction of the manuscript. We appreciate the help and cooperation of the Departments of Polymer Science and Engineering, School of Nano and Advanced Materials Engineering, Gyeongsang National University.

Gyeongsang National University
2010

Jin Kuk Kim
Kaushik Pal

# About the Authors

**Jin Kuk Kim** has been teaching at Polymer Science and Engineering, School of Material Science and Engineering, Gyeongsang National University since 1989. He received his BS and MS degree in Chemical Engineering from Yonsei University, Seoul, South Korea. He holds his Ph.D. degrees in Polymer Engineering from the University of Akron, USA. His main research focus is on wood fiber based solid and microcellular plastics, recycling of waste ground rubber tire, high performance elastomer, lifetime Prediction of elastomers, fuel cell rubbber gasket etc. He is having more than 175 papers in International Journals, more than 40 patents. He has already delivered more than 30 lectures at various Universities and Conferences. He authored and co-authored more than 5 books in the elastomer fields.

**Kaushik Pal** is currently working as a Senior Researcher at Polymer Science and Engineering, School of Nano and Advanced Material Engineering, Gyeongsang National University. He obtained his BE and Ph.D. degrees from Berhampur University and Indian Institute of Technology, Kharagpur, India, respectively. His main area of research is on blending of elastomer and polymer nanocomposites, sensors and actuators, high performance polymer nanocomposites, coatings, rheology, and microcellular foams etc. He received BK-21 fellowship for two years working in Republic of Korea. He already published more than 50 papers in International and National Journals. He serves as a reviewer of some Polymer related International Journal. He has edited one book entitled 'Recent advancement in elastomeric nanocomposites', Springer-Verlag, Germany and authored one book chapters in 'Advances in Polymer Nanocomposites Technology', Nova Science Publishers, USA.

# Contents

# Chapter 1
# Overview of Wood–Plastic Composites and Uses

## 1.1 Introduction

A composite material is made by combining two or more materials to give a unique combination of properties. The above definition is generic and can include metal alloys, plastics, minerals, and wood. Fiber-reinforced composite materials differ from the above materials in that the constituent materials are different at the molecular level and are mechanically separable. In bulk form, the constituent materials work together but remain in their original forms. The final properties of composite materials are better than constituent material properties.

The concept of composites was not invented by human beings; it is found in nature. An example is wood, which is a composite of cellulose fibers in a matrix of natural glue called lignin. The shell of invertebrates, such as snails and oysters, is an example of a composite. Such shells are stronger and tougher than man-made advanced composites. Scientists have found that the fibers taken from a spider's web are stronger than synthetic fibers. In Korea, India, Greece, and other countries, husks or straws mixed with clay have been used to build houses for several 100 years. Mixing husk or sawdust in a clay is an example of a particulate composite and mixing straws in clay is an example of a short fiber composite. These reinforcements are done to improve performance.

Wood–plastic composite (commonly abbreviated as WPC) is a composite material lumber or timber made of recycled plastic and wood wastes. There are also applications in the market, which utilize only virgin raw materials. It's most widespread use is in outdoor deck floors, but it is also used for railings, fences, landscaping timbers, cladding and siding, park benches, molding and trim, window and door frames, and indoor furniture. Manufacturers claim that WPC is more environmentally friendly and requires less maintenance than the alternatives of solid wood treated with preservatives or solid wood of rot-resistant species. Resistant to cracking and splitting, these materials can be moulded with or without simulated wood grain details.

J. K. Kim and K. Pal, *Recent Advances in the Processing of Wood–Plastic Composites*, Engineering Materials, 32, DOI: 10.1007/978-3-642-14877-4_1, 

## 1.2 Why WPC?

For all practical purposes (except in adhesives) a polymer is always reinforced with some kind of fillers. Filled polymer compounds can be used in two main ways. Firstly, to provide comparable performance to unfilled polymer but at a lower costs (diluents or cheapening agents), or secondly, to provide performance well above that achievable using unfilled polymers (using reinforced fillers). However, increasing record oil prices and monomer prices are helping to an upward drive in the cost of fillers. For a polymer technologist a right combination of polymer, filler and additives can provide a wide variety of performance levels in compounds. However, the relationship between performance and cost remains crucial if the application is ever likely to be successful in the market. A simple relationship between modulus and price of polymer is shown in Fig. 1.1 (based on 2006 prices).

The general trend observed in Fig. 1.1 is likely to continue for many more years due to ever increasing prices of monomers due to geo-political factors involved in petroleum exporting countries. From the figure we can observe that commodity plastics like PP, HDPE, HIPS, PS etc. have low modulus but offer advantages like low cost and ease of processability. In order to augment the performance of these commodity plastics, fillers are normally utilized. Several billion pounds of fillers and reinforcements are used annually in plastics industry. The use of fillers in plastics is likely to grow further with development of improved compounding techniques and new coupling agents that permit usage of high filler/reinforcement content. Filled plastics up to 70–75 pph could be common in the future which can have tremendous impact in lowering the usage of petroleum based plastics. The use of fillers by the plastic industry has grown steadily along with the growth in the production of major classes of plastic resins. For example in 1967, the US demand for fillers by the plastic industry was 525,000 tons; filler use had grown to 1,925,000 tons by 1998 (Eckert [1]) and the projected use of fillers by the US plastic industry in 2010 is to 8.5 billion pounds, of which 0.7 billion pounds (8%) was estimated to be bio-based fibers. Although calcium carbonate constitutes the major filler used by weight (66%), it accounted for only 32% ($140 million) of the total value of fillers used in 1998 ($435 million total). Eckert [2] reported average

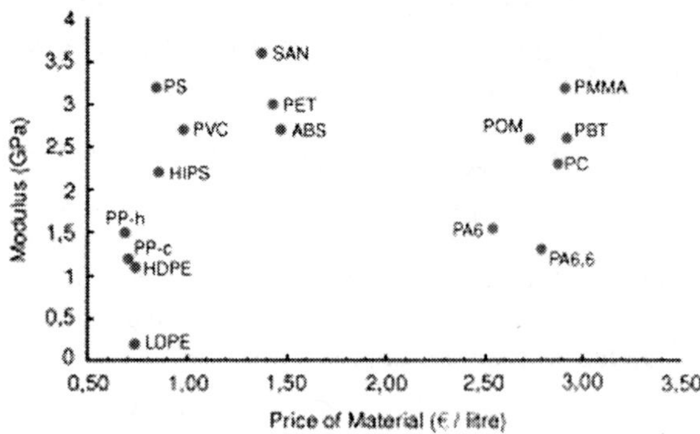


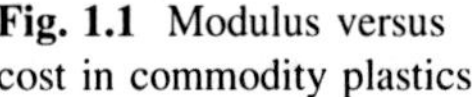
**Fig. 1.1** Modulus versus cost in commodity plastics

per pound prices of commonly used plastic fillers as follows: fiberglass, $0.90, natural fibers other than wood, $0.20, wood fiber, $0.10, and calcium carbonate, $0.70. Most bio-fiber plastic additives are derived from wood. However, other natural fibers, such as flax or wheat straw are finding their way into the fiber/plastic industry. Eckert [2] also summarized major markets for natural fibers in plastic composites as follows, on a weight basis: building products, 70%; other (including marine uses, infrastructure), 13%; industrial consumer, 10%; and automotive, 7%. Although the US annual growth in plastic demand is forecast at approximately 4.5–5.2% per year for 2005–2010, substantially greater growth in the demand for natural fibers is expected. This figure is based on the rate of increase in excess of 50% per year for the period 2000–2005 in the building products area; a significant portion of this growth will be attributed to larger market share for fiber/plastic lumber in residual decking [3]. A smaller, but significant subset of the building products market is also found in vinyl windows [4]. Wood fiber, at weight loading up to 70%, is used in vinyl or vinyl-clad wood window components.

Due to spectacular economic progress witnessed by emerging Asian economies like India, China, Vietnam it is estimated that by the year 2012, 50% of the world's "middle class" will reside in these counties. This represents 5–6 times as much as the US population. If the desire for materials in these growing economies equals to the 'middle class mentality' of US, there will be great demand for new materials. Besides these Asian economies, other emerging markets in South America like Brazil, Mexico, erstwhile Soviet bloc countries in Eastern Europe are also rapidly emerging as hungry consumers that seek new materials.

Based on above facts it is imperative to search for cheaper fillers in plastics besides the traditional filler like calcium carbonate. More ever, the price of Calcium carbonate (filler that is used in bulk in plastics) too is ever increasing due to rapid growth in construction and real estate in Asian economies. Besides the pollution and energy consumption in production of calcium carbonate is also harmful to ecosystem.

One alternative to these mineral fillers is natural fibers and wood. Since prices of plastics have risen sharply over the few past years, adding a natural powder or fiber to plastics provides a cost reduction to the plastic industry (and in some cases increases performance too) but to the agro based industry, this also represents an increased value for agro-based component. The rapidly changing economic and environmental needs of society are putting ever increasing pressures on the forestry industry to do more with less. In practical terms, this means, increasing conversion and efficient use of wood fiber resources, producing more fiber on a shrinking land base, using environmentally friendly processes and technologies, and remaining competitive in the global market place. Within the next decade, composites are expected to constitute the most prominent segment of the board industry. Competition in high volume markets has focused attention on low priced materials that offer a more favorable strength to weight ratio. Compared to other polymeric materials, WPC has the lowest material cost. Wood plastic composites are an attractive alternative because their manufacturing process is highly automated and adaptable to various species and forms of raw materials. Sometimes, wood is not considered as an engineering material because it does not have consistent, predictable, reproducible, continuous, and uniform properties. This might be true for solid wood but is not necessarily true for composites made from wood [5]. Wood flour has been used as filler in synthetic plastics (primarily thermosetting polymers) for decades. The use of wood in thermoplastics is a relatively recent phenomenon spurred by improvements in processing technology, development of suitable chemical coupling agents and economic factors. Advantages such as reductions in operating temperatures, cycle times, and mold shrinkage have also been instrumental in the growth of the fiber/plastic composite industry.

Recent interest in reducing the environmental impact of materials is leading to the development of newer materials or composites that can reduce stress on environment. In light of ever increasing petroleum prices and shortages and pressure for decreasing dependence on petroleum products, there is an increasing interest in maximizing the use of renewable materials. The use of agricultural materials as source of raw materials to plastic industry not only provides a renewable source but could also generate non-food source of economic development for farmers and rural areas. This fact is crucial in countries like India wherein 60% of population reside in villages and who are feeling left out of the economic boom witnessed by the country.

For any long term commercial development there must be a guaranteed long term supply of resources. Mother Nature has bestowed humanity with wide range of natural fibers available in wide range of colors, sizes and shapes. A rough inventory of potential fiber resources available in world is given in the Table 1.1.

From table one can deduce that there can be a perennial supply of fibers but there is a huge need of effective “sustainable exploitation” of these natural resources without causing undue strain on the environment. Sustainable agriculture denoting a balance between the utilization and conservation of agriculture and forest lands is a basic necessity. Unless one particular fiber has some advantage in

**Table 1.1** Inventory of potential fiber sources in the world

| Fiber source | Metric tons (dry) |
|---|---|
| Wood | 1,750,000,000 |
| Straw (wheat, rice, oat, barley etc.) | 1,145,000,000 |
| Stalks (corn, sorghum, cotton) | 970,000,000 |
| Bagasse from sugar cane | 75,000,000 |
| Reeds | 30,000,000 |
| Bamboo | 30,000,000 |
| Cotton staple | 15,000,000 |
| Core based fibers (jute, kenaf, hemp) | 8,000,000 |
| Papyrus | 5,000,000 |
| Bast (jute, kenaf, hemp) | 2,90,000 |
| Cotton linters | 1,000,000 |
| Esparato grass | 500,000 |
| Leaf (sisal, abaca, henequen, pineapple) | 700,000 |
| Sabai grass | 200,000 |
| Total | 4,035,080,000 |

the market it will be replaced with whatever resource available. Market advantage can be based on availability, price and performance of the fiber in WPC. Producers and manufacturers should explore common interests and devise enterprise driven long term plans for development of agro-fiber industry. For example: jute fibers exclusively grown in eastern parts of India and Bangladesh can be effective and cheap source of reinforcements in polymer composites. Other advantages of Jute is low density and its non abrasive nature which makes possible high filling levels resulting in high stiffness, high performance characteristics like modulus, recyclability etc. However it doesn't make economic sense for a WPC manufacturer in Korea to import jute fibers from India and use it as reinforcement. So, the choice of agro based reinforcements depends heavily on the local availability of the fibers.

The data presented in above table suggests that wood will continue to be a major source of agro-based fibers. Another advantage of wood over other agro fibers is its costs. Wood due to its structure and occurrence has higher density than other agro fibers. There are also serious concerns about seasonality of annual crops which requires considerations of harvesting, transporting, separating, drying, storing, cleaning and handling. This seasonality of other agro-fibers will certainly increase their costs and year though availability. Therefore, wood continues to be an ideal choice for agro based polymer composites. Natural fibers can be considered as composites of hollow cellulose fibrils held together by a binder (lignin) in a hemicellulose matrix. The cell wall is inhomogeneous and each type of species of plant has unique fiber characteristics. The general structure of a natural fiber is shown in Fig. 1.2.

Fibers have a complex layered structure consisting of a thin primary wall which is the first layer deposited during cell growth encircling a secondary wall. This secondary wall is in turn made of three distinct layers of which the thickness of the middle layer determines the ultimate mechanical properties of the fiber. The middle layer is a complex and consists of a helically wound cellular microfibrils (typically of 10–30 nm long) from long chain cellulose molecules (made up of 30–100 cellulose molecules) (see Table 1.2).

The angle between the fiber axis and the microfibrils is called the microfibrillar angle. The characteristic value of microfibrillar angle varies from one fiber and another. The smoothness and dexterity of the fiber depends on this microfibrillar

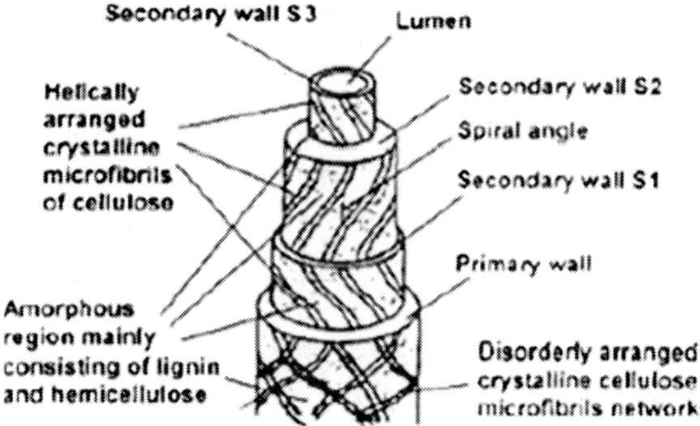


**Fig. 1.2** General structure of natural fibre

**Table 1.2** Typical dimensions of some common lignocellulosic fibers (all dimensions are in mm)

| Type of fiber | Typical length | Average length | Width |
|---|---|---|---|
| Cotton | 10–60 | 18 | 0.02 |
| Flax | 5–60 | 25–30 | 0.012–0.027 |
| Hemp | 5–55 | 20 | 0.025–0.05 |
| Bamboo | 1.5–4 | 2.5 | 0.025–0.04 |
| Esparto | 0.5–2 | 1.5 | 0.013 |
| Cereal (rice, wheat, barley etc.) straw | 1–3.4 | 1.5 | 0.023 |
| Jute | 1.5–5 | 2 | 0.03 |
| Deciduous wood | 1–1.8 | 1.2 | 0.03 |
| Coniferous wood | 3.5–5 | 4.1 | 0.025 |

angle. The amorphous matrix in a cell wall is very complex and consists of hemicellulose, lignin and pectin. The hemicellulose molecules are hydrogen bonded to cellulose and act as cementing matrix between the cellulose microfibrils forming a complex cellulose–hemicellulose network which is the main structural component of the fiber cell. The hydrophobic lignin and pectin components act as coupling agents and increase the stiffness of this cellulose/hemicellulose component. Some of the important natural fibers and their genealogical name is given in the Table 1.3.

**Table 1.3** Typical properties of some natural fibers

| Fiber name | Species | Origin | Tensile strength (MPa) | Youngs' modulus (MPa) | Elongation at break (%) | Density (g/cm$^3$) |
|---|---|---|---|---|---|---|
| Abaca | Musa textiles | Leaf | 400 | 12 | 3–10 | 1.5 |
| Alfa | *Stippa tenacissima* | Grass | 350 | 22 | 5.8 | 0.89 |
| Bamboo | More than 1,200 species | Grass | 140–230 | 11–17 | – | 0.6–1.1 |
| Banana | *Musa indica* | Leaf | 500 | 12 | 5.9 | 1.35 |
| Broom root | *Muhlenbergia macrorua* | Root | | | | |
| Cantala | *Agave cantala* | Leaf | | | | |
| China jute | *Abutilon theophrasti* | Stem | | | | |
| Coir | *Cocus nucifera* | Fruit | 175 | 4–6 | 30 | 1.2 |
| Cotton | *Gossypium* sp. | Seed | 287–597 | 5.5–12.6 | 7–8 | 1.5–1.6 |
| Curaua | *Ananas erectofilius* | Leaf | 500–990 | 11.8 | 3.7–4.3 | 1.4 |
| Date palm | *Peonix dactylifera* | Leaf | 97–196 | 2.5–5.4 | 2–4.5 | 1–1.2 |
| Flax | *Linum usitassimum* | Stem | 345–1,035 | 27.6 | 2.7–3.2 | 1.5 |
| Hemp | *Cannabis sativa* | Stem | 690 | 70 | 1.6 | 1.48 |
| Jute | *Corchorus capsularis* | Stem | 393–773 | 26.5 | 1.5–1.8 | 1.3 |
| Kenaf | *Hibiscus cannabinus* | Stem | 930 | 53 | 1.6 | - |
| Oil palm | *Elaeis guineensis* | Fruit | 248 | 3.2 | 25 | 0.7–1.5 |
| Pineapple | *Ananus comosus* | Leaf | 1.44 | 400–627 | 14.5 | 0.8–1.6 |
| Sisal | *Agave sisilana* | Leaf | 511–635 | 9.4–22 | 2–2.5 | 1.5 |
| Straw (cereal) | More than 500 species | Stalk | | | | |
| Wood | More than 20,000 species | Stem | | | | |

## 1.3 What is Wood?

Wood is a hard, fibrous tissue found in many plants. It has been used for centuries for both fuel and as a construction material. It is an organic material, a natural composite of cellulose fibers (which are strong in tension) embedded in a matrix of lignin which resists compression. Wood may also refer to other plant materials with comparable properties, and to material engineered from wood, or wood chips or fiber (see Fig. 1.3).

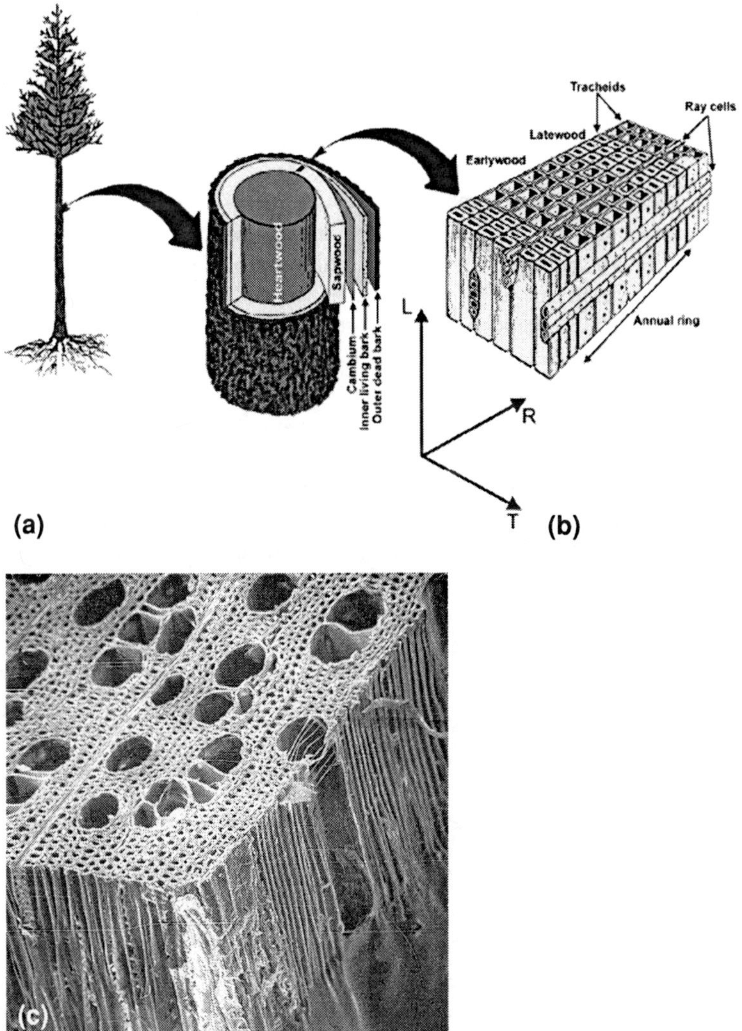


**Fig. 1.3** Schematic illustration of morphological structure of softwood in three principal directions: **a** longitudinal (*L*), **b** radial (*R*) and **c** tangential (*T*)

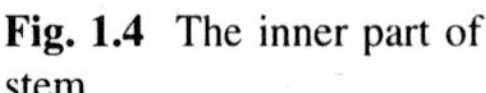

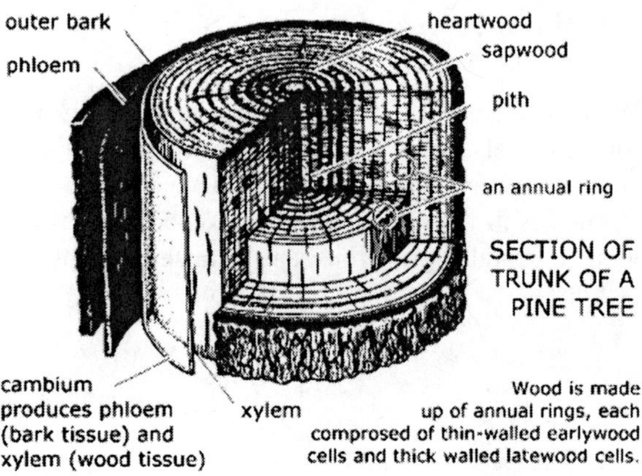


**Fig. 1.4** The inner part of stem

A cross section of a tree trunk reveals different parts of the stem as shown in figure. The outer layer of any tree is of course bark which can be subdivided into the outer dead bark and inner living bark. Adjacent to the inner bark there is the cambium layer which is the growth zone in wood. The inner parts of the stem consist of dead cells in the sapwood and in the heartwood (see Fig. 1.4).

Softwoods consist mainly of longitudinal tracheoidals (90–95%) and a small number of ray cells (5–10%) [6]. The longitudinal tracheids, named here as fibers grow seasonally. In spring and the fibers growth is rapid and are thin walled (termed as early wood fibers) and in summer wherein there will be available of plenty of water and sunshine (latewood fibers) are thick-walled. The average length of fibers varies with species of the wood and also its location. Fibers of trees in tropical or semi-tropical areas have long fibers with thick walls whereas trees in cold locales like scandinavian softwood (Norwegian spruce) fiber is 2–4 mm and the width is 0.02–0.04 mm. Early wood fibers have an average cell wall thickness of 2–4 μm and latewood fibers have an average cell wall thickness of 4–8 μm (see Fig. 1.5).

The difference in cell walls is clearly visible. Both types of wood are made of fibrils of cellulose embedded in a matrix of several other kinds of polymers such as pectin and lignin.

## 1.3.1 Molecular Level

At the molecular level wood fiber is basically made of polymers like cellulose, hemi-cellulose, lignin, pectin in addition to minute percentage of inorganic compounds and other extractives. A brief description of the major constituents of wood is given below.

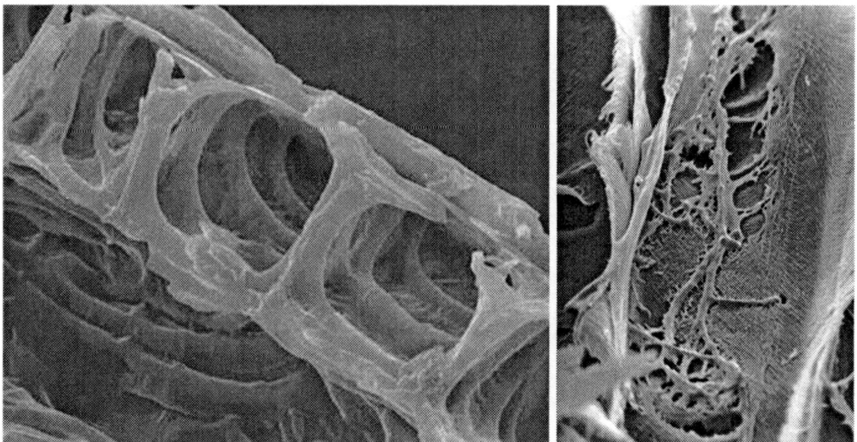

**Fig. 1.5** Plant cellulose cell walls from the Scandinavian softwood (*left*) and a tropical tree (*right*)

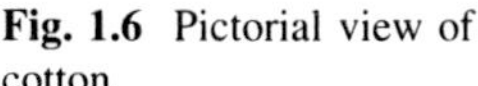

**Fig. 1.6** Pictorial view of cotton

#### 1.3.1.1 Cellulose

Cellulose is the most abundant biopolymer on earth. It is synthesized in plants, trees and grass and even in some varieties of Algae. The most common and abundantly available form of cellulose available to mankind for centuries is cotton (see Fig. 1.6).

Around 4–5% of dry weight of wood consists of cellulose. Cellulose is a linear polymer built of D-glucose units linked together by $\beta$-(1-4)-glycosidic bonds (Fig. 1.7a). Due to the linear nature of cellulose molecules they are capable of

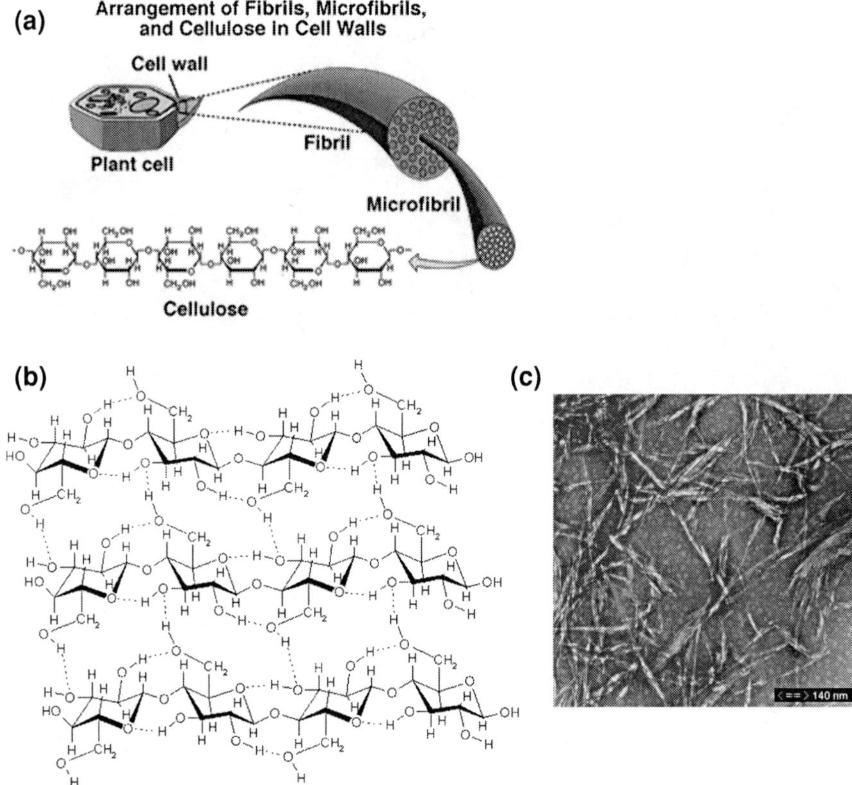


**Fig. 1.7** **a** Structure, **b** schematic and **c** TEM microphotograph of cellulose

forming strong inter and intra-molecular bonds and aggregated bundles of molecules. In scientific literature cellulose bundles have been given many different names such as elementary fibrils, microfibrils, protofibrils etc. Arrangement of these structural aspects in typical wood fiber is shown in Fig. 1.7.

The presence of cellulose in wood fibers can be both useful and detrimental vis-à-vis in wood plastic composites. In living plants the crystalline cellulose has two different unit cells: $I_\alpha$ and $I_\beta$. During pulping, cellulose $I_\alpha$ is converted to $I_\beta$ in alkaline environments. $I_\beta$ is thermodynamically more stable than $I_\alpha$ thus providing higher strength and higher thermal resistance to WPC. A detailed description of effect of crystallinity of cellulose in wood polymer composites will be discussed in the subsequent chapters of this book.

#### 1.3.1.2 Hemicellulose

Hemicelluloses are a group of heterogeneous biopolymers (with degree of polymerization in the range of 200–300) that act as 'structural role' in fiber walls. 20–30% of dry weight of wood consists of hemicelluloses. A wide range of

**Table 1.4** Contents of the main components between softwood and hardwood

| Types | Cellulose | Glucomannan | Xylan | Other polysaccharides | Lignin |
|---|---|---|---|---|---|
| Softwood | 33–42 | 14–20 | 5–11 | 3–9 | 27–32 |
| Hardwood | 38–51 | 1–4 | 14–30 | 2–4 | 21–31 |

hemicellulose biopolymers made of several monomers like mannose, arabinose, xylose, galactose and even glucose are available in wood fibers. The strength and hardness of the wood fibers are intrinsically linked with the percentage and extent of these monomers in the polymers. Besides these monomers, some acidic sugars like galacturonic and glucoronic acids are also present in hemicellulose polymers. In softwood, the hemicellulose galactoglucomannan make up about 16% of dry weight of wood whereas in hardwoods this percentage can be as high as 22–24% (see Table 1.4).

Structures of major constituents of softwood are given in Fig. 1.8.

Many of these components are extracted during treatment of wood fibers in WPC applications. For example pine, spruce and fir wood fibers are usually alkali treated before addition in WPC, this alkali treatment around 50–60% of arabinogalactans consisting of galactose, arabinose and minor components like uronic acids. Extraction of these arabinogalactans prevents rotting and other bacterial attacks of the wood thereby enhancing the lifetime and durability of WPC.

In hard woods the main hemicellulose is Xylan whereas Glycomannan is a minor component (see Fig. 1.9).

Alakli or hot water treatment do not extract either Xylan or Glucomannan in hardwoods thereby hardwood reinforced WPC exhibit higher mechanical properties but with poorer dispersion and higher viscosities thereby making processing difficult.

#### 1.3.1.3 Lignin

Lignins are heterogeneous 3-D polymers that constitute approximately 30% of dry weight of wood. Presence of lignin limits the penetration of water into the wood cells and makes the wood very compact. Lignins are complex polymers based on three monolignols: (a) p-coumaryl alcohol, (b) coniferyl alcohol and (c) sinapyl alcohol (see Fig. 1.10; Table 1.5).

A major factor that governs reinforcing efficiency of fibrous fillers in polymer matrix is fiber dispersion, fiber length distribution, fiber orientation and fiber-matrix adhesion. Most agro fibers are polar and are hydrophilic in nature whereas majority of polymers are non-polar and hydrophobic. These diagonally opposite nature of fibers and polymers can result in effective dispersion of fibers in the matrix thereby giving composites which fall short of their true potential. Clumping or agglomeration must be avoided to produce effective composites. The efficiency of the composite also depends on the extent of stress transferred from the polymer matrix to the fibers. This can be maximized by improving the interactions and adhesion between the two phases and also by maximizing the length of fibers

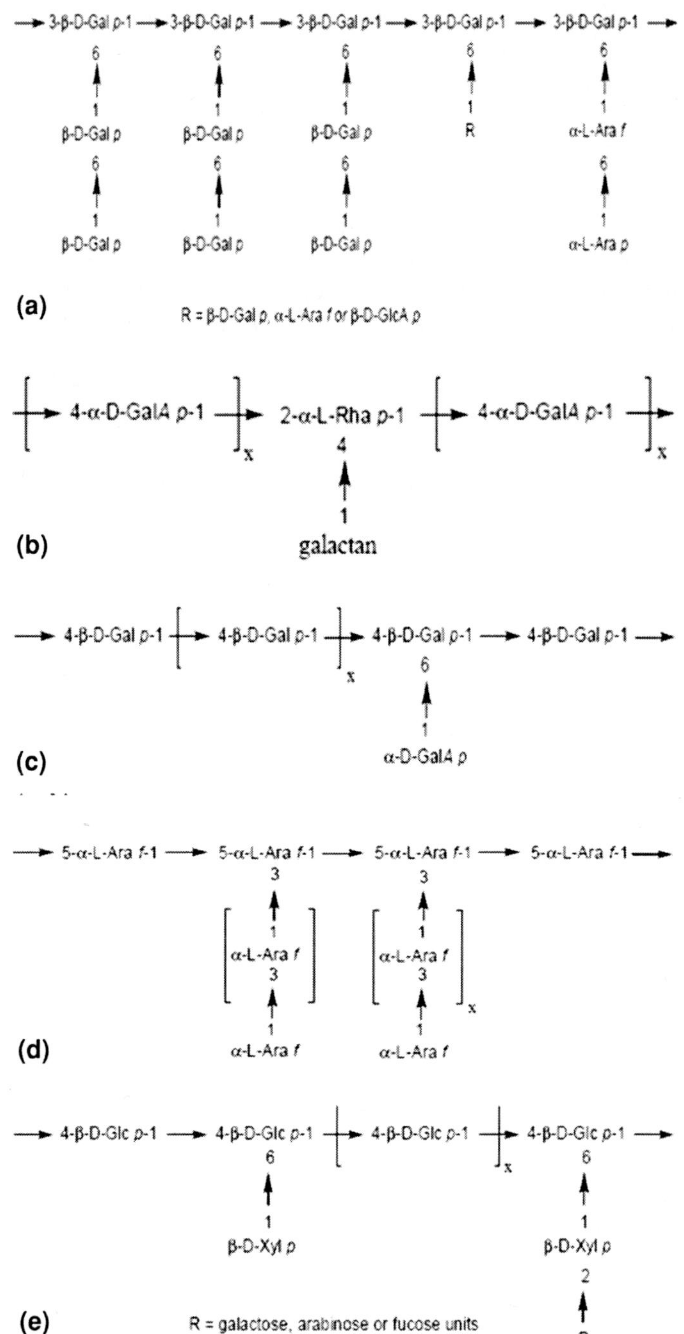


**Fig. 1.8** Some structures of main constituents in soft wood **a** arabinogalactan, **b** rhamnogalactan, **c** pectic galactan, **d** arabinan and **e** xyloglucan

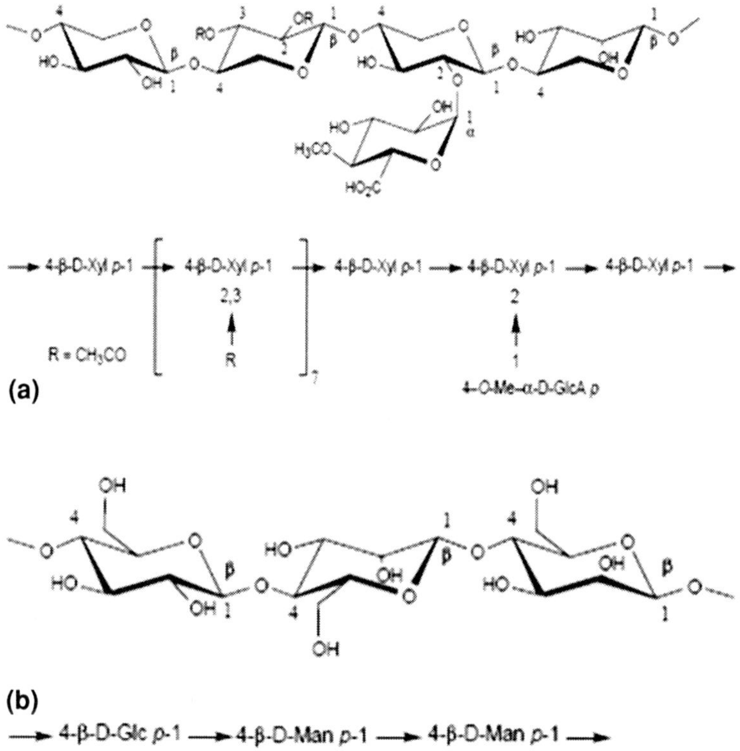

Fig. 1.9 Components of hard woods **a** xylan and **b** glucomannan

Fig. 1.10 Structure of lignin

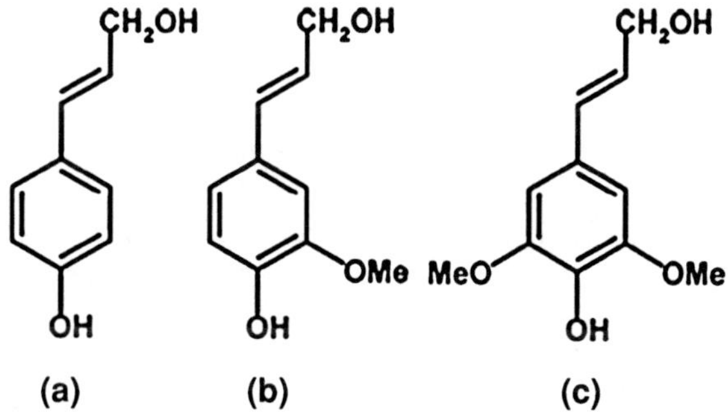

retained in the final composite. Long filamental agro fibers especially jute and coconut fibers can result in higher distribution of fibers in the polymer matrix, however long fibers increase the degree of clumping and agglomeration which ultimately reduces the efficiency of the composite. Many chemical and physical methods can be applied in order to increase the dispersion and fiber-matrix adhesion in WPC. A detailed discussion on various surface treatments will be dealt extensively in subsequent chapters. Besides treating fibers and polymers to

**Table 1.5** Chemical compositions of some common fibers

| Type of fiber | Cellulose | Lignin | Ash | Inorganic |
|---|---|---|---|---|
| Rice stalk | 28–36 | 12–16 | 23–28 | 15–20 |
| Wheat stalk | 29–35 | 16–21 | 26–32 | 4.5–9 |
| Barley stalk | 31–34 | 14–15 | 24–29 | 5–7 |
| Oat stalk | 31–37 | 16–19 | 27–38 | 6–8 |
| Rye stalk | 33–35 | 16–19 | 27–30 | 2–5 |
| Sugar cane | 32–44 | 19–24 | 27–32 | 1.5–5 |
| Bamboo | 26–43 | 21–31 | 15–26 | 1.7–5 |
| Esparato grass | 33–38 | 19–24 | 27–32 | 6–8 |
| Sabai grass | 22 | 23.9 | 6 | – |
| Reed fiber | 44.75 | 22.8 | 20 | 2.9 |
| Flax fiber | 43–47 | 21–23 | 24–26 | 5 |
| Kenaf fiber | 31–39 | 15–19 | 22–23 | 2–5 |
| Jute | 45–53 | 21–26 | 18–21 | 0.6–2 |
| Abaca leaf fiber | 60.8 | 8.8 | 17.3 | 1.1 |
| Sisal leaf fiber | 43–56 | 7–9 | 21–24 | 0.5 |
| Cotton linter seed | 80–85 | – | 0.8–2 | – |
| Coniferous wood fiber | 40–45 | 26–34 | 7–14 | – |
| Deciduous wood fiber | 38–49 | 23–30 | 19–26 | – |

improve mixing efficiency, special compounding techniques and compounding equipment are also available for blending agro fibers and plastics. More detailed discussion on equipment and machinery will be dealt in subsequent chapters.

## 1.4 WPC in Automotive Applications: A Case Study

Many governments and NGOs like Green Peace are increasingly addressing the environmental impact of automobiles. As a consequence of this awareness governments are pushing for more stringent regulations and legislations to preserve and protect the quality of environments for future generations. To overcome this problem at source (at materials level) chemical industries are designing innovative materials benign of any chemical reactions with minimal energy consumption, the so called 'green chemistry'. Composite industries are also seeking more environmentally friendly materials. There is an ever increasing interest in novel biodegradable and bio-based composites based on either natural polymers or renewable fibers. The combination of interesting mechanical and physical properties together with sustainable character has triggered various activities in polymer composite industry named as 'green-composites'. Wood or natural fiber based composites are a consequence of this philosophy. However inherent drawbacks such as tendency of the reinforcements to form aggregates during processing, low thermal stability, abysmally low resistance to moisture and very large variations (even in individual plants of same species) greatly reduce the potential of plant fibers to be used as reinforcements. Besides these the inherent moisture absorption capacity of plant fibers leading to swelling and presence of voids at the interface results in poor mechanical properties and dimensional stability. Usage of coupling agents and treatment of fibers with hydrophobic materials can reduce the moisture absorption but it can never be eliminated. So, natural fiber composites are specifically used in situations which have low probability of making contact with water and moisture. Besides this, natural fibers are also known to suffer wide degradation when subjected to atmosphere and sunlight for extended periods of time. This also makes the usage of WPC in open spaces very unreliable.

Automobiles use substantial amounts of plastics especially in their interiors. A typical example is a dashboard. The following figure shows the evolution of dashboard from the famous Ford Model T to the modern day BMW E-class passenger car (see Fig. 1.11).

The sea change in the dash board can be clearly observed. Most of the components of these dash boards are made of plastic, mostly ABS. Schematic diagram of a typical automobile dash board is shown in the Fig. 1.12.

Most of these components can be in fact replaced by natural fiber composites which not only reduce the weight of car, thereby saving fuel consumption, but also improve the aesthetics of car interior. Besides this fuel savings, a light weight car

**Fig. 1.11** Pictorial views of car **a** car body and **b** dash board parts

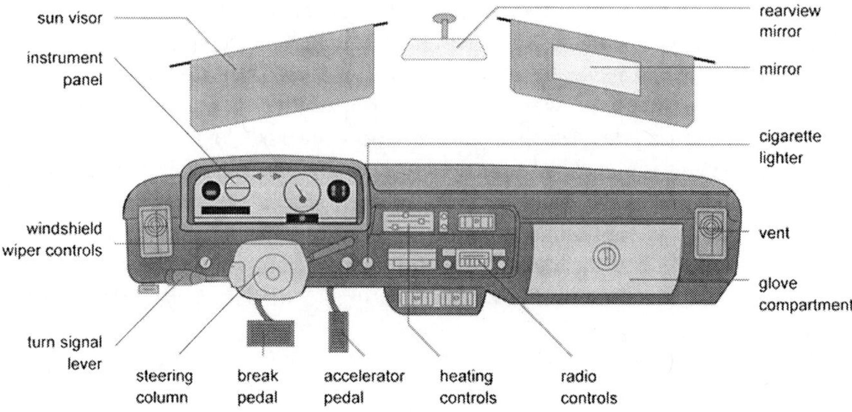


**Fig. 1.12** Interior parts of the car

also has higher safety considerations. A study by CAFÉ in 2001, Corporate Average Fuel Economy standards showed that lighter cars accounted for lower deaths in case of accidents and roll overs when compared to bulkier cars. Besides trim parts in dash boards, door panels, parceled shelves, seat cushions, backrests and cabin linings, WPC has potential to replace fiberglass and even some steel components. One of the major automotive applications is the utility of plant fibers for thermo-acoustic insulation. Recent studies have also been reported on the effect of flax fibers in car disk brakes to replace environmentally harmful asbestos fibers. So far there are very few exterior parts made from plant fiber composites. The following diagram shows the WPC market in USA (see Fig. 1.13).

From the figure it can be observed that automotives are extensive consumers of WPC closely followed by construction and marine applications. This tremendous share of WPC in automotives is not only a result of greater concern for environmental protection but also the recent European Guideline 2000/53/EG administered by EU which stipulates that 85% of the weight of vehicle has to be recyclable. This recyclable percentage is legislated to increase to 95% in 2015 of which 85% should

**Fig. 1.13** Sector wise consumption of wood plastic composites used in 2008

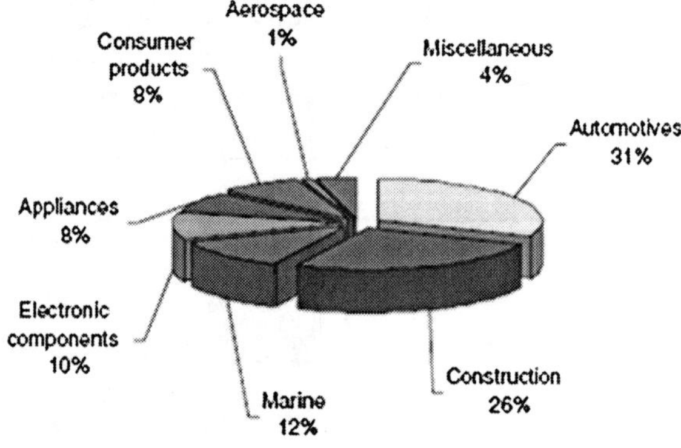

be recycled or reused by mechanical recycling and 10% through energy recovery or thermal recycling. Consequently, the usage of WPC in automobiles is projected to grow further, substantially increasing the demand of WPC. From a technical point of view WPC are natural choice with these bio-based composites will enhance mechanical strength and acoustic performance, reduce material weight, energy/fuel consumption and processing time, lower production costs, improve passenger safety and provide shatter proof performance under extreme temperatures (especially low temperatures) and improve biodegradability of auto interior parts. A perfect example of this is Mercedes E-class cars which showed a reduction of 20% in weight of the car coupled with improved mechanical properties which is an important factor for passenger protection in the event of accidents.

Daimler–Benz was the earliest companies which has been exploring the idea of replacing glass fibers with plant fibers in automotive components since 1991. Mercedes used jute based door panels in its E-class range of cars in 1996. In 2000, Diamler–Chrysler began using plant fibers for their vehicle production. The bast fibers used by them impart weight savings of between 10–30% and corresponding cost savings without compromising vehicles performance. After the pioneering work of Diamler companies, many German auto manufacturers like Mercedes, Volkswagen, Audi, BMW, Ford and Opel now use WPCs in various applications. Following chart shows the increasing natural fiber content in German automobiles (see Fig. 1.14).

In 2000, Audi launched A2 midrange car in market with door trim panels made of polyurethane reinforced with mixed flax and sisal fiber mat. The latest industry buzz is research of Daimler Chrysler in flax reinforced polyester composites for exterior applications (see Table 1.6).

A further 5 kg could be easily integrated in other vehicle interior parts. With worldwide automobile production of 58 million vehicles (cars, STVs etc.) the

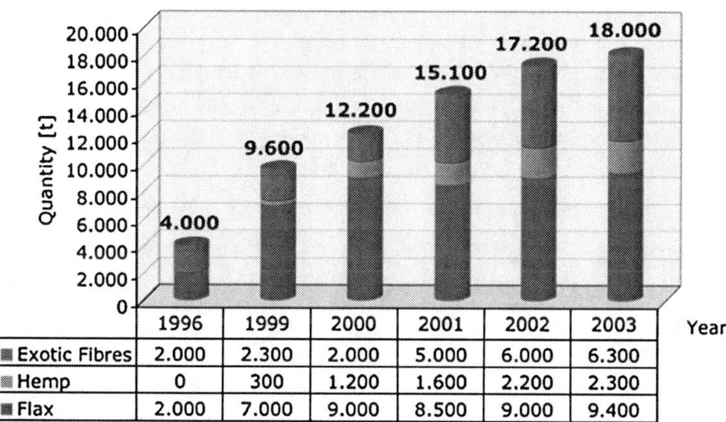


| | 1996 | 1999 | 2000 | 2001 | 2002 | 2003 |
|---|---|---|---|---|---|---|
| Exotic Fibres | 2.000 | 2.300 | 2.000 | 5.000 | 6.000 | 6.300 |
| Hemp | 0 | 300 | 1.200 | 1.600 | 2.200 | 2.300 |
| Flax | 2.000 | 7.000 | 9.000 | 8.500 | 9.000 | 9.400 |

**Fig. 1.14** Use of natural fibres (except cotton and wood) for composites in the German automotive industry 1996–2003

**Table 1.6** Typical amounts of plant fibers used for different applications in cars

| Component | Weight (kg) |
|---|---|
| Front door linens | 1.2–1.8 |
| Rear door linens | 0.8–1.5 |
| Boot linens | 1.5–2.5 |
| Parcel shelves | 1.8–2 |
| Seat backs | 1.6–2 |
| Headliners | 2.5–3 |

potential for natural fibers is in the range of 175,000 to 350,000 tons per annum. A recent estimate by FAO and Common Fund for Commodities (CFC) in US suggests a market potential of 35,000 tons for fibers such as Kenaf, Jute, hemp, flax and sisal in automotives with an annual growth rate expected to be in the range of 10–14%.

## 1.5 Impact of WPC on Humanity

Whether you are aware of it or not, WPC play an important part in your life. WPC versatility allows them to be used in everything from car parts to doll parts, from soft drink bottles to the refrigerators they are stored in. From the car you drive to work into the television you watch at home, WPC help make your life easier and better. So how is it that WPC have become so widely used? How did WPC become the material of choice for so many varied applications?

The simple answer is that WPC can provide the things consumers want and need at economical costs. WPC have the unique capability to be manufactured to meet very specific functional needs for consumers. So maybe there's another question that's relevant: what do I want? Regardless of how you answer this question, WPC can probably satisfy your needs.

If a product is made of WPC, there's a reason. And chances are the reason has everything to do with helping you, the consumer, get what you want: health; safety; performance; and value. WPC help make these things possible.

### *1.5.1 Shopping*

Just consider the changes we have seen in the grocery store in recent years: research is being carried out in WPC bio-composite wraps which help keep meat fresh while protecting it from the poking and prodding fingers of your fellow shoppers; biodegradable WPC bottles mean you can actually lift an economy-size bottle of juice and should you accidentally drop that bottle, it is shatter-resistant. In each case, WPC help make your life easier, healthier and safer.

### 1.5.2 Grocery Cart versus Dent-Resistant Body Panel

Wood–plastic composite also help you get maximum value from some of the big-ticket items you buy. WPC help make portable phones and computers that really are portable added with aesthetic look. They help major appliances—like refrigerators or dishwashers—resist corrosion, last longer and operate more efficiently. WPC car fenders and body panels resist dings, so you can cruise the grocery store parking lot with confidence.

### 1.5.3 Packaging

Modern packaging—such as natural fiber reinforced WPC pouches and wraps–helps keep food fresh and free of contamination. That means the resources that went into producing that food aren't wasted. It's the same thing once you get the food home: plastic wraps and re-sealable containers keep your leftovers protected—much to the chagrin of kids everywhere. In fact, packaging experts have estimated that each pound of plastic packaging can reduce food waste by up to 1.7 lb.

Wood–plastic composite can also help you bring home more product with less packaging. For example, just 1.6 lb of WPC can deliver 1,300 oz—roughly 10 gallons—of a beverage such as juice, soda or water. You would need 3 lb of aluminum to bring home the same amount of product, 8 lb of steel or over 40 lb of glass. Not only do plastic bags require less total energy to produce than paper bags, they conserve fuel in shipping. It takes seven trucks to carry the same number of paper bags as fits in one truckload of plastic bags. WPC make packaging more efficient, which ultimately conserves resources.

### 1.5.4 Light Weighting

Wood–plastic composite engineers are always working to do even more with less material. Since 1977, the 2-l plastic soft drink bottle has gone from weighing 68 g to just 47 g today, representing a 31% reduction per bottle. That saved more than 180 million pounds of packaging in 2006 for just 2-l soft drink bottles. The 1-gallon plastic milk jug has undergone a similar reduction, weighing 30% less than what it did 20 years ago.

Doing more with less helps conserve resources in another way. It helps save energy. In fact, WPC can play a significant role in energy conservation. Just look at the decision you're asked to make at the grocery store checkout: "Paper or plastic?" Plastic bag manufacture generates less greenhouse gas and uses less fresh water than does paper bag manufacture. Not only do plastic bags require less total production energy to produce than paper bags, they conserve fuel in shipping. It takes seven trucks to carry the same number of paper bags as fits in one truckload of plastic bags.

### 1.5.5 Home Construction

Wood–plastic composite also help to conserve energy in your home. Vinyl siding and windows help cut energy consumption and lower heating and cooling bills. Furthermore, the US Department of Energy estimates that use of plastic foam insulation in homes and buildings each year could save over 60 million barrels of oil over other kinds of insulation.

The same principles apply in appliances such as refrigerators and air conditioners. Plastic parts and insulation have helped to improve their energy efficiency by 30–50% since the early 1970s. Again, this energy savings helps reduces your heating and cooling bills. And appliances run more quietly than earlier designs that used other materials.

## 1.6 End Life of WPC

The lifetime of WPC can be shown in the Fig. 1.15.

End life of plastics is a huge environmental issue. Due to the growing awareness regarding plastics recycling among people coupled with stringent legislations and rules it is imperative for WPC manufacturer to think about recycling too. Recycling can be divided into three categories: mechanical recycling, feedstock recycling and source reduction.

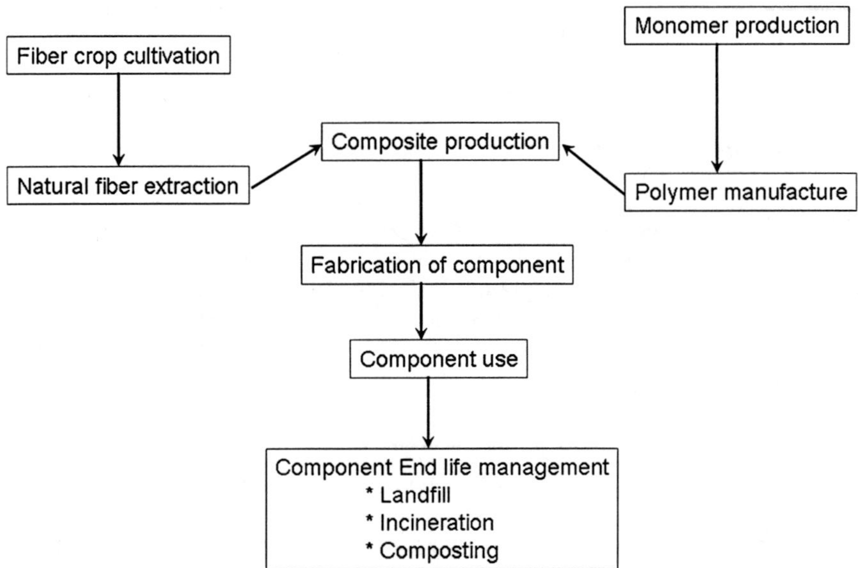


**Fig. 1.15** Life cycle analysis of agro based polymer composite

### 1.6.1 Mechanical Recycling

Recycling of post-consumer plastics packaging began in the early 1980s as a result of state level bottle deposit programs, which produced a consistent supply of returned PETE bottles. With the addition of HDPE milk jug recycling in the late 1980s, plastics recycling have grown steadily but relative to competing packaging materials.

Roughly 60% of the US populations—about 148 million people—have access to a plastics recycling program. The two common forms of collection are: curbside collection—where consumers place designated plastics in a special bin to be picked up by a public or private hauling company (approximately 8,550 communities participate in curbside recycling) and drop-off centers—where consumers take their recyclables to a centrally located facility (12,000). Most curbside programs collect more than one type of plastic resin; usually both PETE and HDPE. Once collected, the plastics are delivered to a material recovery facility (MRF) or handler for sorting into single resin streams to increase product value. The sorted plastics are then baled to reduce shipping costs to reclaimers.

Reclamation is the next step where the plastics are chopped into flakes, washed to remove contaminants and sold to end users to manufacture new products such as bottles, containers, clothing, carpet, plastic lumber, etc. The number of companies handling and reclaiming post-consumer plastics today is over five times greater than in 1986, growing from 310 companies to 1,677 in 1999. The number of end users for recycled plastics continues to grow. The federal and state government as well as many major corporations now supports market growth through purchasing preference policies.

Early in the 1990s, concern over the perceived reduction of landfill capacity spurred efforts by legislators to mandate the use of recycled materials. Mandates, as a means of expanding markets, can be troubling. Mandates may fail to take health, safety and performance attributes into account. Mandates distort the economic decisions and can lead to sub optimal financial results. Moreover, they are unable to acknowledge the life cycle benefits of alternatives to the environment, such as the efficient use of energy and natural resources.

### 1.6.2 Feedstock Recycling

Pyrolysis involves heating plastics in the absence or near absence of oxygen to break down the long polymer chains into small molecules. Under mild conditions polyolefins can yield petroleum-like oil. Special conditions can yield monomers such as ethylene and propylene. Some gasification processes yield syngas (mixtures of hydrogen and carbon monoxide are called synthesis gas, or syngas). In contrast to pyrolysis, combustion is an oxidative process that generates heat, carbon dioxide, and water.

Chemical recycling is a special case where condensation polymers such as PET or nylon are chemically reacted to form starting materials.

### *1.6.3 Source Reduction*

Source reduction is gaining more attention as an important resource conservation and solid waste management option. Source reduction, often called "waste prevention" is defined as "activities to reduce the amount of material in products and packaging before that material enters the municipal solid waste management system."

Source reduction activities reduce the consumption of resources at the point of generation. In general, source reduction activities include:

- Redesigning products or packages so as to reduce the quantity of the materials used, by substituting lighter materials for heavier ones or lengthening the life of products to postpone disposal.
- Using packaging that reduces the amount of damage or spoilage to the product.
- Reducing amounts of products or packages used through modification of current practices by processors and consumers.
- Reusing products or packages already manufactured.
- Managing non-product organic wastes (food wastes, yard trimmings) through backyard composting or other on-site alternatives to disposal.

## References


1. Eckert, C.: In: Proceedings of the Fifth International Conference on Woodfiber-plastic Composites. Proceedings No. 7263, Forest Products Society, Madison (1999)
2. Eckert, C.: In: Proceedings of the Conference on Progress in Woodfibre-plastic Composites Conference, University of Toronto (2000)
3. Smith, P.M.: US Wood Fiber-plastic composite decking market. In: Proceedings of Sixth International Conference on Woodfiber-Plastic Composites. Forest Products Society, Madison, pp. 13–17 (2001)
4. Cannon, C.: In: Proceedings of the Fifth International Conference on Woodfiber-Plastic Composites, Proceedings No. 7263, Forest Products Society, Madison (1999)
5. Rowell, R.M.: Chemical Modification of Wood: A review, Commonwealth Forestry Bureau, Oxford, England, **6**(12), 363–382 (1993)
6. Fengel, D., Grossner, D.: In: Holz, morphologie und Eigenschaften in Ullmanns Encycklopedie der technischen chemie, vol. 12, Weinhein-Verlag Chemie, pp. 669–679 (1976)

# Chapter 2
# Surface Modifications in WPC with Pre-Treatment Methods

## 2.1 Introduction

Compared with conventional and mineral filler reinforced thermoplastic products, wood–polymer composites (WPC) have many advantages such as high specific strength and modulus, low cost, low density, and low friction during compounding. Unlike wood, WPC have excellent dimensional stability under moisture exposure [1, 2] and better fungi and termite resistance [3, 4]. For WPC, one of the most attractive features is that it can help recycle thermoplastic and wood wastes. Therefore, WPC have developed quickly in the last three decades [5]. However, polar wood fiber and non-polar thermoplastics and commodity plastics like PE, PP are not compatible, thus resulting in poor adhesion resulting in weak interface. [6–8].

Although the use of wood based fillers is not as popular as the use of mineral or inorganic fillers, wood derived fillers have several advantages over traditional fillers and reinforcing materials: low density, flexibility during processing with no harm to environment and equipment, acceptable specific strength properties and lower cost per volume basis. Initial studies about the influence of filler and its size on the mechanical and physical properties of wood fiber reinforced thermoplastics indicate a decrease in elongation at break coupled with decrease in modulus and impact strength with increasing concentration of wood fiber independent of the wood fiber size. However, some reports show that the size and dimensionality of the wood filler have an adverse affect on the tensile modulus and tensile strength with increasing aspect ratio of the wood filler (wood fiber rather than wood powder) showing better properties. However the general consensus is that wood powder–plastic composites have low properties due to inherent poor compatibility between hydrophobic thermoplastic and hydrophilic cellulose fibers.

In order to increase the compatibility two kinds of treatments are in practice. First is to make the hydrophobic thermoplastic more hydrophilic by plasma or some sort of high energy surface treatments which impart oxygen donating groups like carboxyl, hydroxyl, etc., groups. However, this process is more complicated

J. K. Kim and K. Pal, *Recent Advances in the Processing of Wood–Plastic Composites*, Engineering Materials, 32, DOI: 10.1007/978-3-642-14877-4_2,

and involves higher capital costs. Consequently this method is confined only to academic and scientific research communities.

Second and the most common method are chemical modification and treatment of wood. Some of these methods are rather simple and involve chemical treatments with alkalis, permanganate, isocyanate and peroxide treatments. These treatments not only remove the unwanted lignin content from the surface of wood fibers but also increase the oxygen content on their surface thereby increasing the polymer–wood fiber interactions. This increased oxygen content helps in considerable enhancement of tensile properties of the composites.

## 2.2 Polymers Used in WPC

Many types of polymers namely thermoplastics, plastics, thermosets and elastomers can be used for fabrication of WPC. But the main consideration which has to be taken in account is the operating temperature. Due to the limitations imposed by the processing of wood (operating temperatures should not exceed 200°C), only plastics or polymers which can be processed at temperatures lower than 200°C can be used for WPC fabrication. The following are the six main types of plastics that can be used in WPC.

### *2.2.1 High Density Polyethylene (HDPE)*

HDPE is used for many packaging applications because it provides excellent moisture barrier properties and chemical resistance. However, HDPE, like all types of polyethylene, is limited to those food packaging applications that do not require an oxygen or $CO_2$ barrier. In film form, HDPE is used in snack food packages and cereal box liners; in blow-molded bottle form, for milk and non-carbonated beverage bottles; and in injection-molded tub form, for packaging margarine, whipped toppings and deli foods. Because HDPE has good chemical resistance, it is used for packaging many household as well as industrial chemicals such as detergents, bleach and acids. General uses of HDPE include injection-molded beverage cases, bread trays as well as films for grocery sacks and bottles for beverages and household chemicals.

### *2.2.2 Low Density Polyethylene (LDPE)*

LDPE is predominantly used in film applications due to its toughness, flexibility and transparency. LDPE has a low melting point making it popular for use in applications where heat sealing is necessary. Typically, LDPE is used to manufacture flexible films such as those used for dry cleaned garment bags and produce bags. LDPE is also used to manufacture some flexible lids and bottles, and it is widely used in wire and cable applications for its stable electrical properties and processing characteristics.

### 2.2.3 Polyvinyl Chloride (PVC)

PVC has excellent transparency, chemical resistance, long term stability, good weatherability and stable electrical properties. Vinyl products can be broadly divided into rigid and flexible materials. Rigid applications are concentrated in construction markets, which include pipe and fittings, siding, rigid flooring and windows. PVC's success in pipe and fittings can be attributed to its resistance to most chemicals, imperviousness to attack by bacteria or micro-organisms, corrosion resistance and strength. Flexible vinyl is used in wire and cable sheathing, insulation, film and sheet, flexible floor coverings, synthetic leather products, coatings, blood bags, and medical tubing.

In near future PVC will not be fittable from the environmental view because it produces the toxic gases during burning.

### 2.2.4 Polypropylene (PP)

PP has excellent chemical resistance and is commonly used in packaging. It has a high melting point, making it ideal for hot fill liquids. PP is found in everything from flexible and rigid packaging to fibers for fabrics and carpets and large molded parts for automotive and consumer products. Like other plastics, polypropylene has excellent resistance to water and to salt and acid solutions that are destructive to metals. Typical applications include ketchup bottles, yogurt containers, medicine bottles, pancake syrup bottles and automobile battery casings.

### 2.2.5 Polystyrene (PS)

PS is a versatile plastic that can be rigid or foamed. General purpose polystyrene is clear, hard and brittle. Its clarity allows it to be used when transparency is important, as in medical and food packaging, in laboratory ware, and in certain electronic uses. Expandable polystyrene (EPS) is commonly extruded into sheet for thermoforming into trays for meats, fish and cheeses and into containers such as egg crates. EPS is also directly formed into cups and tubs for dry foods such as dehydrated soups. Both foamed sheet and molded tubs are used extensively in take-out restaurants for their lightweight, stiffness and excellent thermal insulation.

## 2.3 What is Coupling Agent?

Another method of increasing the interactions of wood fibers and polymers is the usage of coupling agents. Coupling agents can be defined as the substances that are used in small quantities to treat a surface so that bonding occurs between it and other surfaces between wood and plastic in the present scenario. Coupling agents can be subdivided into two broad categories: bonding agents and surfactants (also known as surface active agents). Some other chemicals such as compatibilizers and dispersing agents are also added into the category of coupling agents, but it is wrong. For example stearic acid and metallic salts like zinc stearate are used to improve the dispersibility of wood fibers in the matrix. They primarily act by reducing the interfacial energy at the wood fiber–plastic matrix thereby reducing aggregation of wood fibers facilitating the formation of new interfaces. Certain compatibilizers are also used in WPC which provided compatibility between the plastic and wood fiber through reduction of interfacial tension. Certain compatibilizers like acetic anhydride and methyl isocyanate are mono functional reactants act by lowering the surface energy of the wood fiber and make it non-polar making it more similar to the plastic. However it must be mentioned that certain di or multifunctional compatibilizers like maleated polypropylene (PP-MA), maleated styrene–ethylene/butylenes–styrene (SEBS-g-MA) and styrene-maleic anhydride (SMA) act both as dispersing aids (by lowering the surface energy) and also compatibilizers by formation of chemical bonds between MA and –OH groups present on the wood fiber surface. (A more detailed discussion on this is given in subsequent sections). Thus a functional distinction between bonding agents, coupling agents, compatibilizers and dispersing agents should be noticed. But the general consensus in academia and industry is that a coupling agent act as bonding agent which links the wood fibers and the thermoplastic polymers by one or more of the following mechanisms: covalent bonding, polymer chain entanglement and strong secondary interactions (such as hydrogen bonding).

Meyer [9] in 1968 was probably the first person who suggested that by using a coupling agent (which he called as cross linking agent) one can improve the mechanical properties of wood–plastic composites. Subsequently Gaylord [6] in 1972 had a patent granted which uses Maleic anhydride as a coupling agent in cellulose–polyethylene and cellulose–PVC in the presence of free radical initiator (like AIBN). However, little attention was focused on the usage of coupling agents in WPC until 1980s. However, in late 1980s a series of patents were awarded to Coran and Patel [7], Geottler [8], Nakamura [10], Woodhams [11] on the application of isocyanate and maleic anhydride coupling agents in WPC. Xanthos [12] was probably the first to use silane coupling agent in WPC. He used methacryloxypropyltrimethoxy-silane (A-174) the industrially used silane coupling agent in silica reinforced TPEs to treat wood flour. Tremendous increases in mechanical and physical properties were achieved by this treatment with better dispersion. Subsequently Schneider and Brebner [13] in 1985 used a combination of methacryloxypropyltrimethoxy-silane (A-174) and propylene oxide to achieve even better properties in wood flour-PP composites.

The next big step in the utility of coupling agents in WPC field was introduced by Klason [1] and co-workers. In their pioneering work, Klason successfully introduced Epolene, E-43 a kind of low molecular weight maleic anhydride grafted PP as a coupling agent in thermo-mechanical pulp (TMP) and isotactic polypropylene composites.

The most important issues in chemical coupling agents include:

- Coupling agent absorption and fixation.
- Optimum compounding conditions.
- Characterization of interface and coupling agent distribution.
- Interfacial bonding mechanisms and coupling agent performance.
- Searching for newer coupling agents with greater graft efficiency.
- Durability of the coupling agents, etc.

Based on the coating and grafting techniques, coupling agent treatment of WPC can be classified into three main categories:

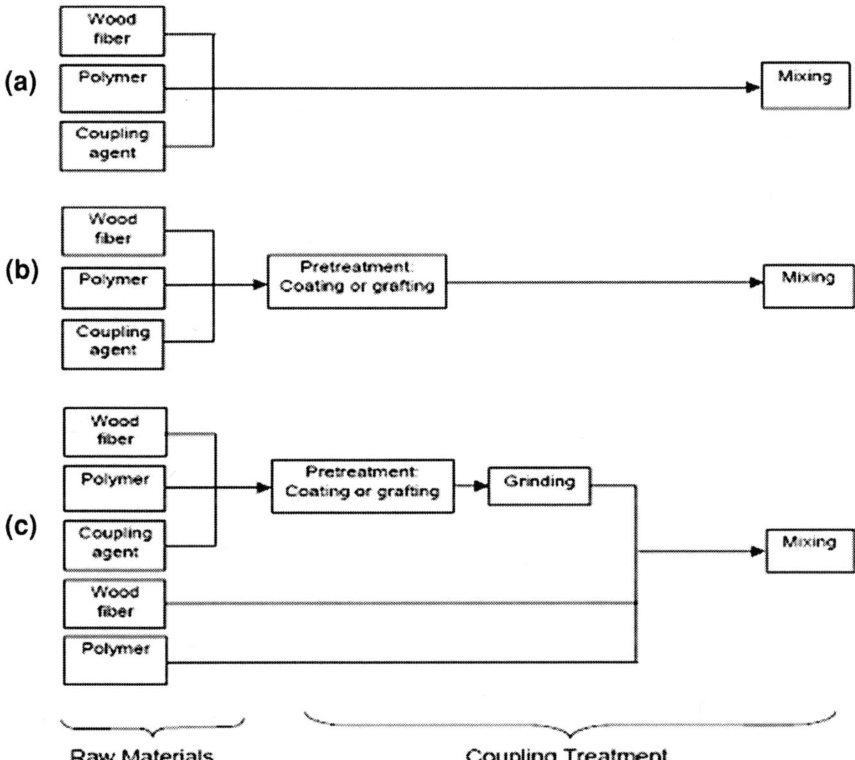


**Fig. 2.1** Schematic representations of three basic coupling treatments of WPC

(a) Coupling agents can be directly coated on wood fiber and polymer during mixing, the so-called one-step process.
(b) Two step process, involving coating or grafting of coupling agents on wood fibers before mixing.
(c) In the third process, a part of polymer and wood fibers are treated with a coupling agent and made as a master batch which can be eventually used in WPC fabrication by suitable dilution (Fig. 2.1).

## 2.4 Classification of Coupling Agents

There are over 40 coupling agents that have been used in WPC of which only a handful are inorganic majority of them being organic compounds. But the most common coupling agents are organic–inorganic hybrids like silane and titanate coupling agents, though pure chemists treat them as organo-metallics. Some of the common coupling agents used in WPC is tabulated.

### *2.4.1 Acrylates*

Glycidyl methacrylate (GMA) and hydroxy methyl methacrylate (HEMA) are one of the oldest and most commonly used coupling agents in WPC. Commercially these coupling agents are available as terpolymers like ethylene-(acrylic ester)-glycidyl methacrylate terpolymers (Fig. 2.2)

The effectiveness of these coupling agents can be further increased by utilizing it in conjunction with maleic anhydride, which gives rise to oligoesterified wood. This oligoesterification of wood results in tremendous increase in mechanical properties due to chemical comptabiliztion with the polymer matrix. The typical oligoesterification reaction of wood is shown in Fig. 2.3

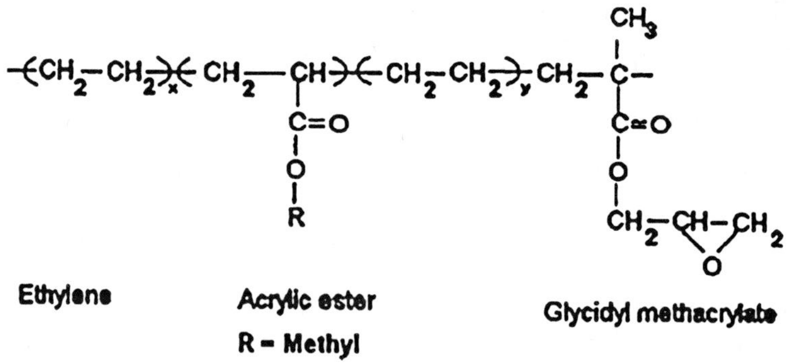

**Fig. 2.2** Structure of commercial glycidyl methacrylate terpolymers

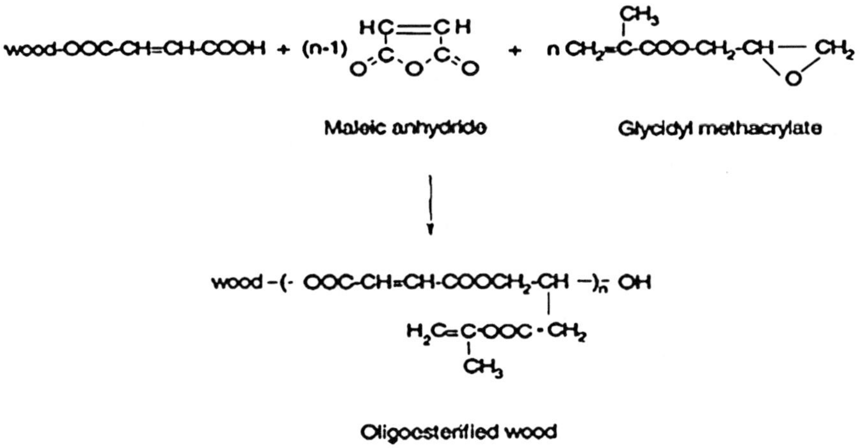

Maleic anhydride

Glycidyl methacrylate

Oligoesterified wood

**Fig. 2.3** Schematic diagram of oligoesterification of wood

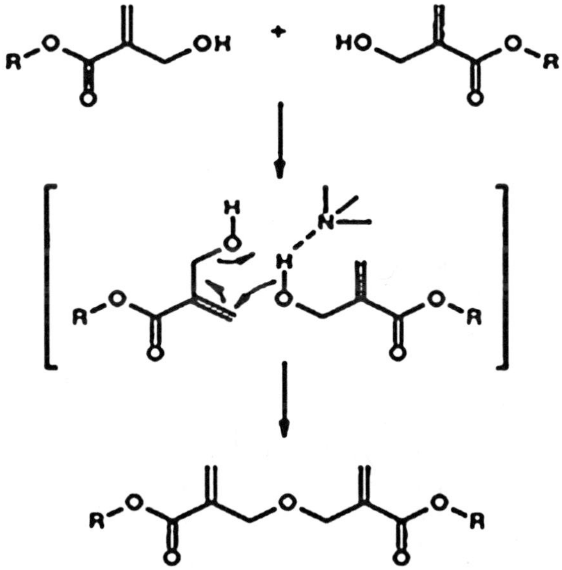

**Fig. 2.4** Base catalyzed ether formation in HEMA treated wood fibers

In case of HEMA, ether moieties are formed of wood by the following base catalyzed reaction (Fig. 2.4).

## 2.4.2 Amides and Imides

*N*, *N*′-*m*-phenylene bismaleicimide (BMI) treatment of wood powder can perform as interface modifiers in WPC. BMI can act as exceptional coupling agent if the cellulose content of the fibers is very high. That is why BMI is the de-facto

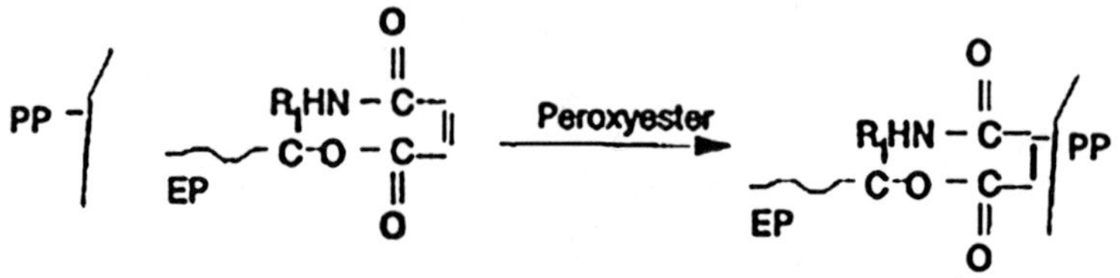

Fig. 2.5 Schematic diagram of BMI treatment of PP

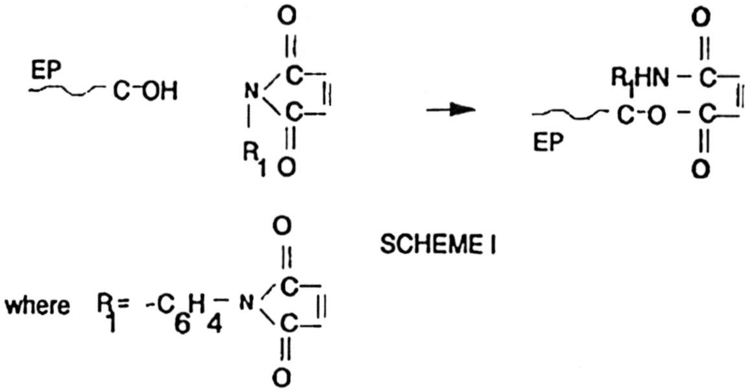

Fig. 2.6 Schematic representation of BMI treatment of wood fiber

coupling agent in wood or natural or biocomposites based on for cellulose whiskers, cellulose fiber and cellulose fibers. BMI can also be used directly to modify the polymer matrix (especially in PP). The typical procedure of this is treatment of powdered PP powder with 5% *m*-phenylene bismaelicimide in presence of 3 wt% peroxyester catalyst in acetone at 135°C for 3 h in nitrogen or argon atmosphere. BMI acts by "substituent electron effect" wherein nitrogen functionalities get involved which will substantially modify the characteristic binding pattern significantly by the following mechanism (Fig. 2.5).

Conversely BMI treatment of pulp or wood fiber can also be carried out in isopropanol at room temperature and dried in vacuum at 60°C for 24 h before use. This treatment substantially increases the N/C atomic ratio of the fiber by the following mechanism (Fig. 2.6).

### 2.4.3 Anhydrides

Anhydrides acetic anhydride (AA), alkyl succinic anhydride (ASA), succinic anhydride (SA), phthalic anhydride (PHA) and maleic anhydride (MA) have been the most widely used and efficient way for coupling wood fibers with polymer matrix. Treatment of wood fiber with anhydrides falls in the category of esterification of the hydroxyl groups present on wood into an ester group. Chemical

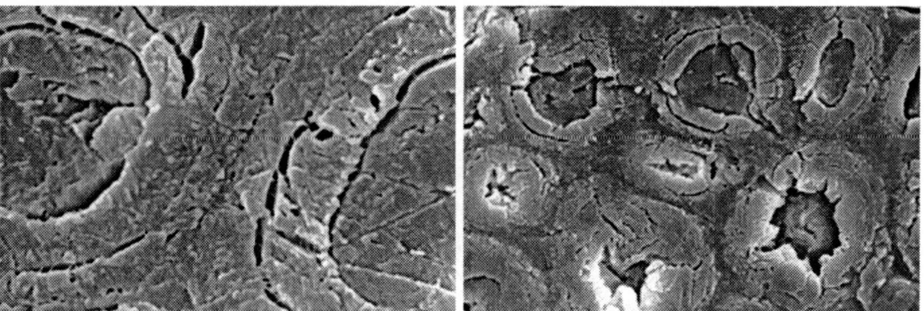

**Fig. 2.7** Bulking of cell wall after acetic anhydride treatment of sisal fibers

**Fig. 2.8** Scheme for esterification of OH groups of wood fibers with an acid donor

modification of wood fibers with acetic anhydride substitutes the cell wall polymer hydroxyl groups with acetyl groups, modifying the properties of these polymers so that they become hydrophobic. In addition, such modification also causes bulking of cell wall proportional to the extent of substitution or weight percent gain (WPG) a renders the wood fibers less susceptible to biological decay (Fig. 2.7).

In woods this is usually done through nucleophilic addition of an organic acid anhydride or acid chloride. Esterification is schematically shown in Fig. 2.8.

The advantage of acid catalyst is to accelerate this reaction by making the carbon of the carboxylic group more positive to increase the nucleophile attraction. The usage of fatty acid chlorides also eliminates the utility of solvents. Besides anhydrides and acid chlorides, fatty acids like stearic and acetic acid can also be used which not only increases the dimensional stability, preservation and wood–plastic compatibility but also results in tremendous increase in performance characteristics.

#### 2.4.3.1 Treatment with Acetic Acid

A variety of anhydrides with different catalysts and solvents have been used common of which are acetic anhydride, maleic anhydride, phthalic anhydride, succinic anhydride, etc., and shown in the Fig. 2.9.

Of all the available anhydrides, maleic anhydride is possibly the widest used. Reaction of anhydrides with hydroxyl groups of wood fibers forms mono or diesters as shown by the schematic diagram (Fig. 2.10).

XPS or ESCA studies of anhydride treated wood fiber are shown in Figs. 2.11 and 2.12.

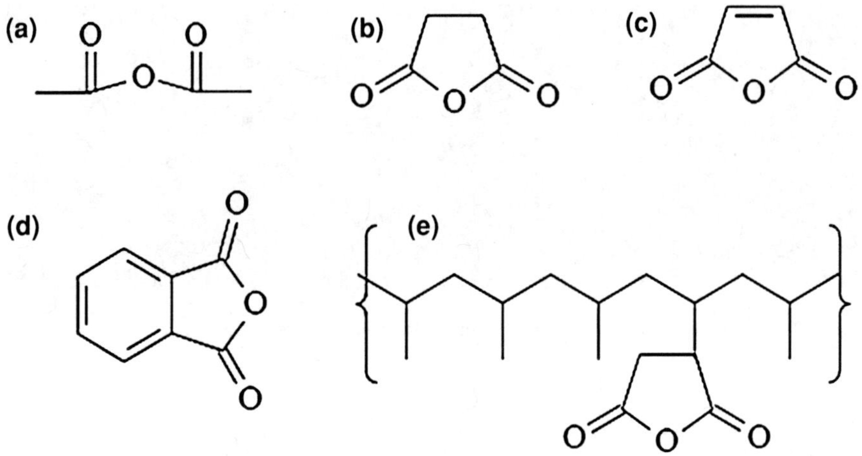

Fig. 2.9 Chemical structures of common anhydrides used to modify wood fibers: **a** acetic anhydride, **b** succinic anhydride, **c** maleic anhydride, **d** phthalic anhydride, **e** PP-g-MA

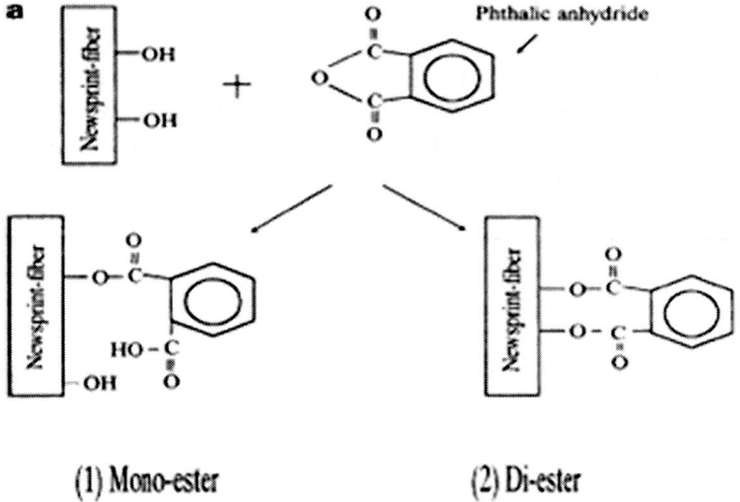


**Fig. 2.10** Reaction between newsprint fibers with phthalic anhydride to form esters

### *2.4.4 Chlorotriazines and its Derivatives*

Chlorotriazines and its derivatives like 2-diallyamino 4,6-dichloro-*s*-triazine (AACA), 2-octylamino 4,6-dichloro-*s*-triazine (OACA), methacrylic acid, 3-(4, 6-dichloro-s-triazine-2-yl) amino)propyl ester (MAA-CAAPE) can also be used as coupling agents in WPC. The main advantage of treating with chloroazines and its derivatives is that they can also act as photostabilizers in WPC. This is very important in HDPE based WPCs, since HDPE and wood fibers undergoes

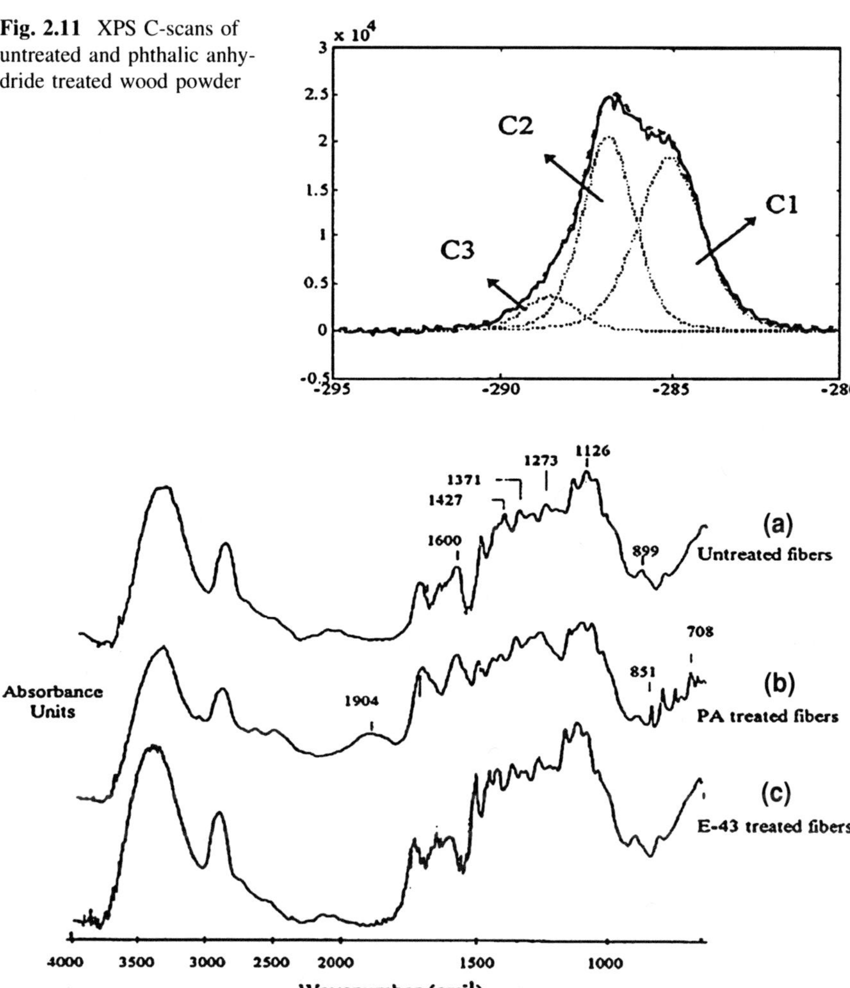


**Fig. 2.11** XPS C-scans of untreated and phthalic anhydride treated wood powder

**Fig. 2.12** FTIR spectra of untreated and phthalic anhydride treated fibers

extensive photodegradation and changes its appearance, the so-called photobleaching reaction. The reaction pathway is phenoxy quinone redox cycle as shown in Fig. 2.13.

By treating wood with chloroazines, hindered amine light stabilizer moieties (HALS) are grafted onto the wood fiber surface thereby effectively shielding the phenoxy-quinone redox cycle. Triazines and diesters are the most common type of cycloazines and can be grafted by dissolving these in appropriate solvent and by steeping steam exploded or alkali treated wood fibers for 24 h at 60°C, drained, dried in vacuum oven (Fig. 2.14).

Of the many cycloazines, tertiary triazines show superior ant-degradation properties. Although the exact reasons for this superiority are not clear, chemists

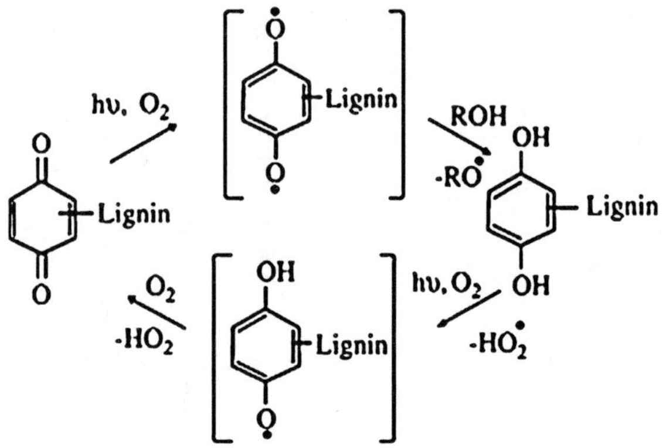

**Fig. 2.13** Phenoxy quinone redox cycle involved in photoyellowing or photobleaching in HDPE based WPC composites

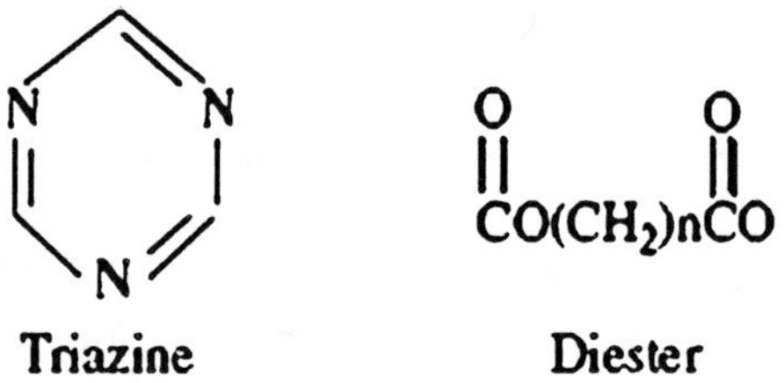

**Fig. 2.14** Triazine and diester backbones of cycloazine coupling agents used in WPC

attribute it to the fact that the latter is more sterically hindered than diesters which make it more stable. However, sufficient care must be taken regarding the purity of tertiary triazines, since even minute amounts of secondary triazine impurities can cause rapid increase in the carbonyl absorbance (around 1,733 $cm^{-1}$) which is a result of scavenging free radical tendency of secondary amines that leads to chromophore formation and results in phot-yellowing (Fig. 2.15).

In case of diesters, higher molecular weight diesters are more efficient than their lower molecular weight counterparts, especially in long term. In short periods

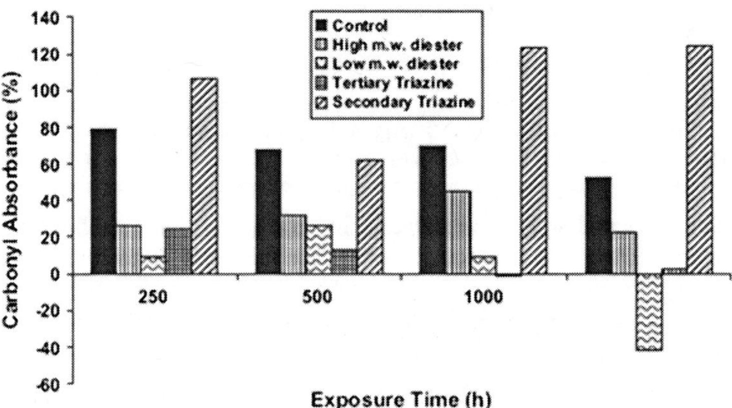


**Fig. 2.15** Effect of triazines and esters in carbonyl absorbance with exposure time

of exposure, lower molecular weight diesters can be very effective since they are upwardly mobile and migrate to the surface very easily. However the disadvantage is they are unable to overcome surface evaporation and maintain their auto-protection mechanism over long periods of exposure time.

### 2.4.5 Epoxides

Treating wood fibers with epoxides like Butylene Oxide (BO) and Propylene Oxide (PO) fall in the category of etherification of the hydroxyl groups of the wood surface. This treatment is invariably preceded by a pre-treatment step with alkali like NaOH. This step is important since the charged intermediates formed during alkali treatment allow faster nucleophilic addition of epoxides as shown in the Fig. 2.16.

Other oxygen donating epoxides like epichlorohydrin can also be used. Epochlorohydrin has a strained ring (3, 4 member) oxygen containing structure that creates electron withdrawal from adjacent carbons. But excessive concentrations of epichlorohydrins in conjunction with phenols (pentachlorophenols) epoxidize, leading to formation of poorly dispersed wood. Etherification of wood by allyl bromide also improves the adhesion between two wooden surfaces.

Oven dried wood fiber or wood pellets are charged into a stainless steel reaction vessel with a mixture of PO or BO and triethylamine (95 + 5, vol:vol) at 120°C and 150 psi nitrogen pressure for 4 h in case of butylenes oxide or for 1 h in case of propylene oxide. The treated solution was drained off, dried under a fumed hood, oven dried.

### 2.4.6 Isocyanates

Isocyanates like ethyl isocyanate (EIC) and hexamethylene diisocyanate (HMDIC), toluene 2,4-diisocyanate (TDIC) are commodity coupling agents, easily available and cheap. The commonest of these are toluene diisocyanate and phenyl diisocyanate (Fig. 2.17).

Other heavy chain diisocyanates such as PMDI, polymeric diphenylmethane diisocyanate has become one of the most important wood binders, primarily for

$$\text{Cell—OH} + \text{NaOH} \longrightarrow \text{Cell—O}^{\ominus}\text{Na}^{\oplus} + H_2O$$

$$\text{Cell—O}^{\ominus}\text{Na}^{\oplus} + \text{Cl—R} \longrightarrow \text{Cell—O—R} + \text{NaCl}$$

**Fig. 2.16** Reaction scheme showing alkali catalyzed reaction of hydroxyl group of cellulose with alkali halide

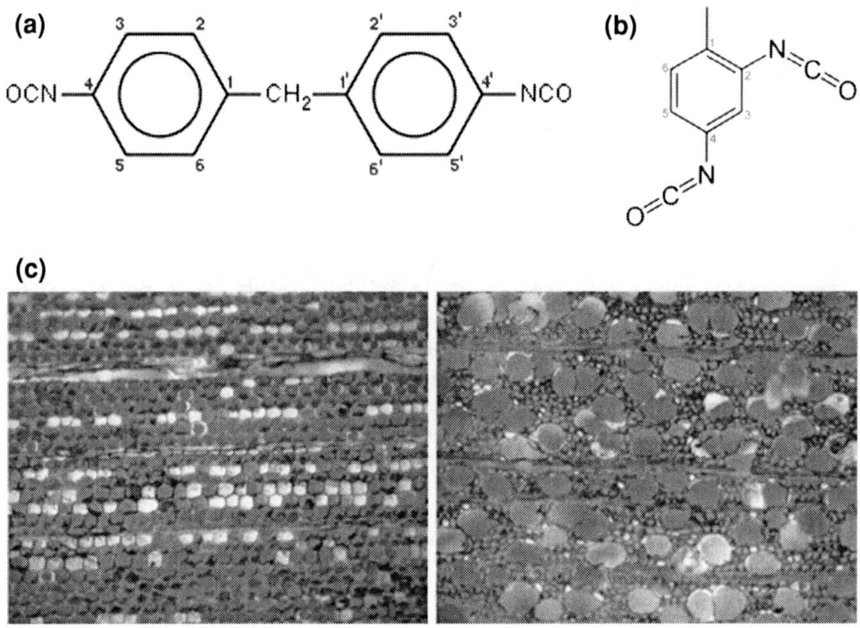

Fig. 2.17 Fluorescence micrographs of **a** MDI, **b** TDI and **c** impregnated pine wood

bonding of oriented strand boards (OSB) and similar wood particulate composites. The market domination of PMDI can be measured by the fact in 2001, PMDI accounted for ~21% of total resin solids used in OSB manufacturing industry in North America. Commercially utilized PMDI is a mixture of MDI oligomers and monomers with average functionality of 2.8 and resin isocyanate content of 30–32%. The molecular structure of commercially available PMDI resins is shown in Fig. 2.18

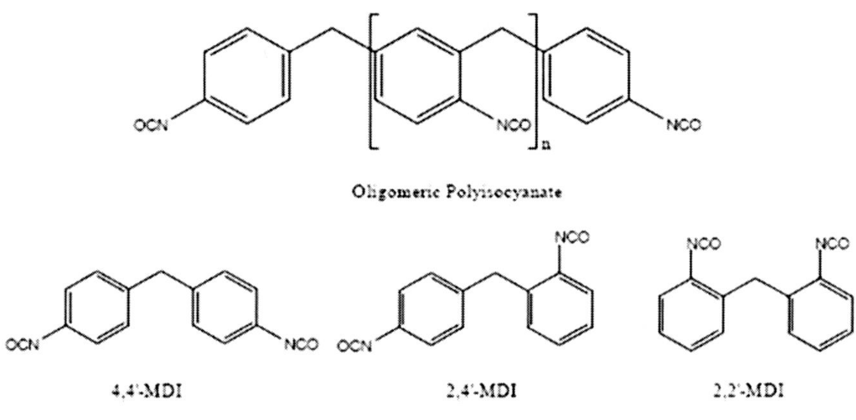

Fig. 2.18 Molecular structure of commercially available PMDI

The main advantage of PMDI over other commodity resins is:

- Cure at high moisture conditions.
- Superior mechanical properties.
- Excellent weather durability.
- No formaldehyde emissions (important for indoor applications).

However PMDI resins are relatively expensive than commodity resins like PF or CI resins, but posses advantages like faster production rates, lower resin loadings and lower press temperatures. The lower press temperature is especially attractive considering the high costs of energy in recent times. However, PMDI has certain disadvantages like:

- Excessive adhesion of panels to the walls and press platens that necessitates usage of releasing agents.
- Toxicity of isocyanates.
- Special storage requirements to avoid premature curing especially in hot and humid atmospheres.

Grafting of isocyanates on wood fibers is very simple and involves treating either steam exploded or NaOH treated wood fibers with either benzyl isocyanate and phenyl isocyanate dissolved in anhydrous acetone under constant stirring or

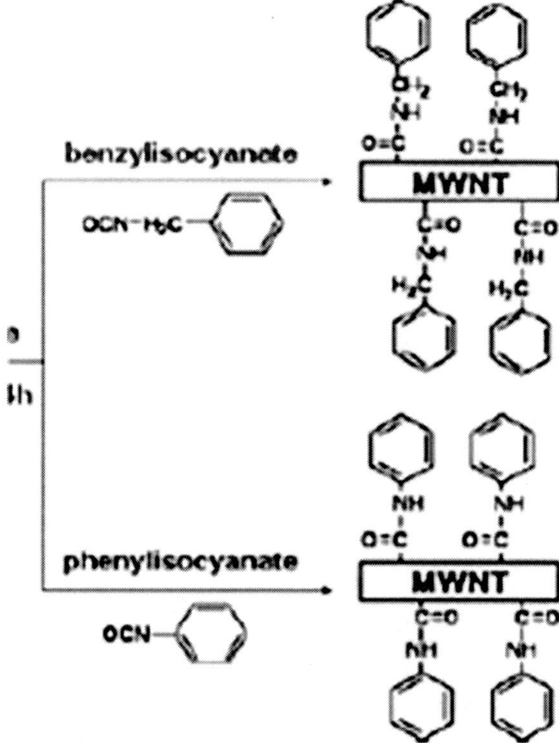


**Fig. 2.19** Scheme for preparing isocyanate treated wood fibers

ultrasonic conditions in a dry nitrogen atmosphere at 50°C for 24 h followed by drying in a vacuum oven at 50°C for 24 h (Fig. 2.19).

However sufficient care must be taken to avoid premature curing since NCO radicals of isocyanates are highly susceptible to attack by atmospheric moisture and may result in premature curing.

However in case of cellulose fibers or wood fibers three different schemes of isocyanate grafting have been proposed (Fig. 2.20).

Above scheme shows that the superficial OH groups on the surface of wood fibers can interact effectively with the reactive moieties such as –NCO yielding urethane linkages thereby the isocyanates are chemically bonded to the wood fiber surface. Treating wood fibers with isocyanates decreases the polarity and hydrophilicity of wood fiber therefore increasing the compatibility of treated fibers with the polymer matrix.

The efficiency of isocyanate grafted or isocyanate treated wood fibers is considerably increased in presence of secondary compatibilizing agents like PP-g-MA or SEBS-g-MA or any maleic anhydride donating species. However in this case, there may be two competitive reactions occurring as shown in Fig. 2.21.

The –MA group reacts with –OH groups of cellulose forming an ester bond, thereby making the PP-g-MA matrix chemically bonded to the wood fibers. However, it is well known that –NCO group readily reacts with the anhydride group forming imide linkages. However this imide linkage formation occurs only at higher temperature (especially during compounding step like in extrusion or

$$\mathrm{R{-}N{=}C{=}O + HO{-}Cell \longrightarrow R{-}HN{-}\overset{\overset{\displaystyle O}{\|}}{C}{-}O{-}Cell}$$

Fig. 2.20 Scheme of –OH group interactions with NCO– radicals of isocyanates

Scheme I

Scheme II

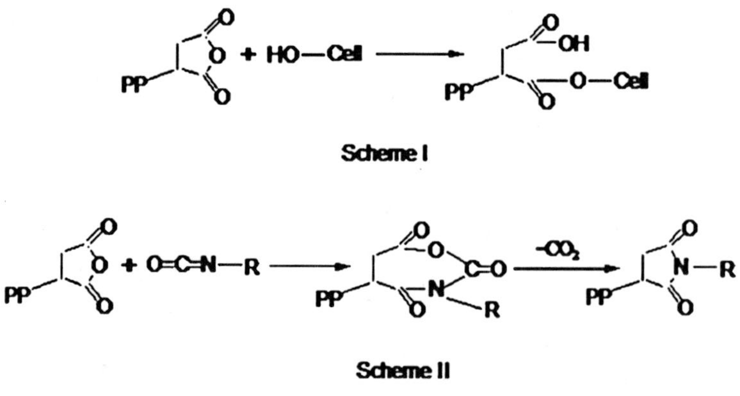

Fig. 2.21 Reaction of PP-g-MA with isocyanate treated wood fiber

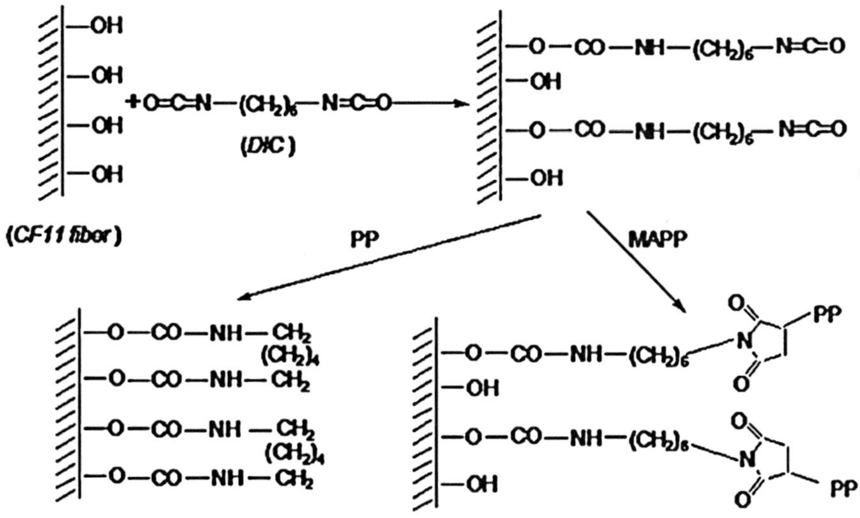

**Fig. 2.22** Reaction of isocyanate in PP-g-MA/PP/wood fiber composites

injection molding). This causes substantial hardening of the extrudate thereby making it extremely difficult to process. Addition of isocyanate treated wood fiber in conjunction with PP-g-MA also leads to formation of other side reactions like biuret formation and isocyanate dimerization. In order to avoid these unwanted reactions, the residence time of extrudate during extrusion and injection molding should be minimized (Fig. 2.22).

### *2.4.7 Grafting of Monomers*

A graft copolymer consists of a polymeric backbone with covalently linked polymeric side chains like acrylonitrile (AN), butyl acrylate (BA), epoxypropyl methacrylate (EPMA), methacrylic acid (MAA), methyl emthacrylate (MMA), styrene (St) and vinyl monomers.

### *2.4.8 Polymers and Copolymers*

Other technique to improve surface compatibilization is based on using surfactant type of structure like ethyl-vinyl acrylate (EVAc), maleated polymers like maleated polypropylene (PP-MA), maleated polyethylene (PE-MA), maleated EPDM (EPDM-MA), phenylene bismaleicimide modified polypropylene (BPP), polymethacrylic acid (PMAA), polyvinyl acetate (PVAC), phenol–formaldehyde resin (PF Resin), Styrene–ethylene–butylene–styrene–maleic anhydride (SEBS-g-MA), styrene–maleic anhydride (SMA). In this technique molecules bearing one or more polar end groups which are capable of reacting with OH groups, in order to graft long hydrophobic 'hairs'(single terminal reactive moiety) or 'bridges' (multiple reactive moieties) on the wood or natural fiber surface, capable of protecting their surface from water uptake and to make it compatible with non-polar matrices like polyolefins. This procedure is shown schematically in Fig. 2.23. *N*-Octyl-4,6-dichloro-*s*-triazine is on such surfactant used to treat fibers in order to limit water uptake of fiber based unsaturated polyester composites. Typically, the wood fibers or natural fibers are steeped in surfactant dissolved acetone for 3 min. Other surfactants like alkyl ketene dimmers, alkenyl succenic anhydride (shown in Fig. 2.23) and stearic acid have also been used in order to compatibilize surface of natural fibers with polypropylene.

The action of surfactants showed that both dispersive and polar components of the surface energy as well as the acid–base work of adhesion decreased drastically when the fibers had been treated with surfactants. Another important aspect is the 'synergy' between surfactant treated natural fiber and polyolefins when compatibilizers like pol(styrene-maleic anhydride) or poly(propylene-graft-maleic

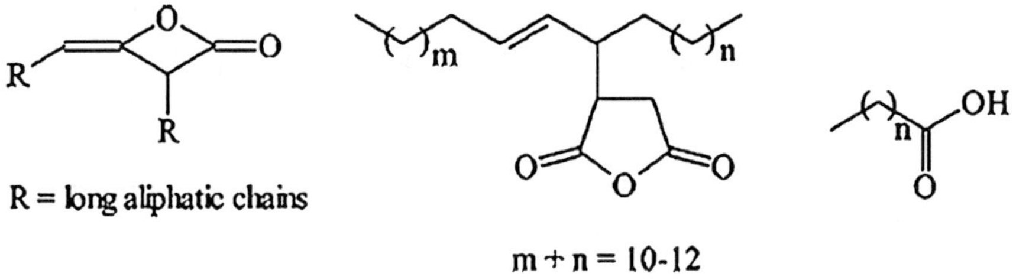

R = long aliphatic chains

$m + n = 10\text{-}12$

**Fig. 2.23** Some typical surfactants used to treat natural fibers

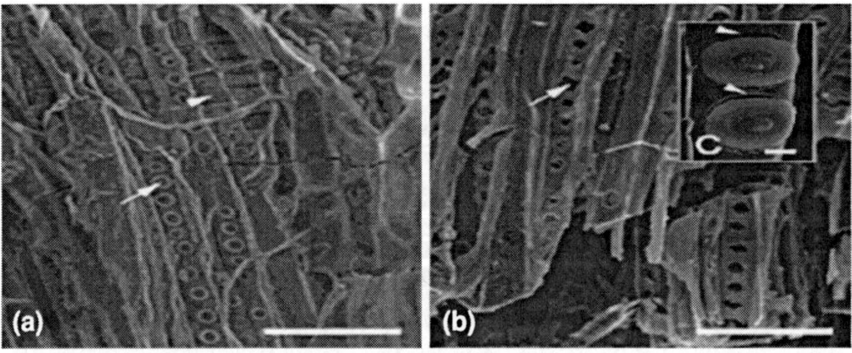


**Fig. 2.24** SEM micrograph of **a** silica impregnated wood and **b** tracheids with bordered pots (insets picture with higher magnification)

anhydride) are used. Studies have shown a considerable decrease in polar contribution of surface energy from 43 to 21 mJ/m$^2$. The treated fibers not only showed excellent dispersion but also significant increase in properties like tensile strength and modulus due to enhanced stress transfer arising from better stress transfer to the dispersed fibers due to decrease in polar nature of the fibers.

Recently some studies have also used silica impregnation technique to reduce the surface energy of wood particles, thereby increasing the polarity of the wood. This silica impregnation not only results in excellent dispersion but also good performance properties resulting from the presence of additional filler namely, nano-silica. A SEM of silica impregnated wood fiber is shown in Fig. 2.24.

This silica impregnation can be achieved by commercially available silanes like methyltrimethoxy silanes and can be carried out in commonly occurring solvent mediums. The advantage of this technique is that unreacted silane can also act as surfactant wherein head of a surfactant molecule is a polar group like Si–OH and hence this ionizes in water and hence is hydrophilic. The tail is a covalently bonded hydrocarbon polymer (methyltrimethoxyl oligomer) and hence hates the high dielectric constant medium of water and is hydrophobic. A schematic representation of this surfactant action is shown in Figs. 2.25 and 2.26.

## 2.5 Physical Treatments

Wood fibers and its other forms like wood flour, wood pellets can be modified by physical methods. Physical treatments do not change the chemical composition of the wood but considerable changes in structural and surface properties of the fiber occur thereby influencing the mechanical bonding with the polymer matrix. In case of wood fibers physical methods such as stretching, calendaring, thermo-treatment and production of yarn is carried in order to ease the handling and transportation of the fibers. Other forms of physical treatment like electric corona or cold plasma discharge, electron beam irradiation is another way of physical treatment. These treatments are the most easy and versatile method of physical treatment to increase and improve surface oxidation and activation. In case of wood, electron beam

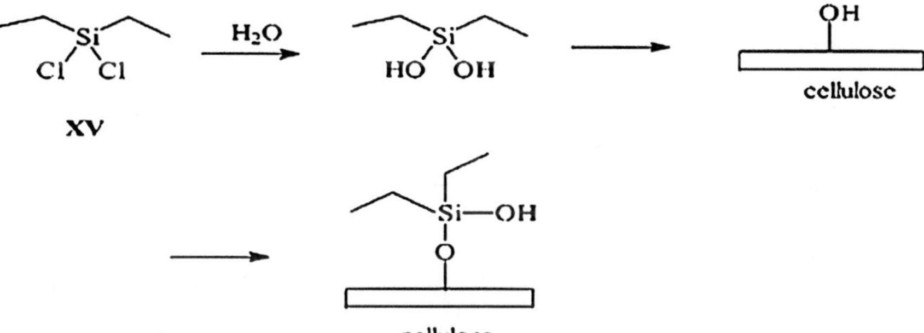


**Fig. 2.25** Grafting of silane surfactant on wood fibers

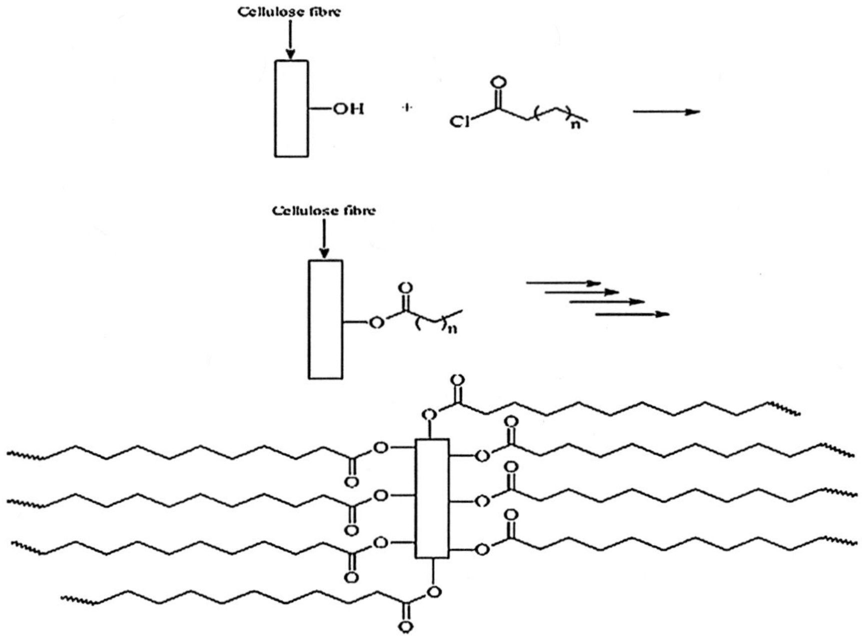

Fig. 2.26 Reaction of surfactant (*n*-octyl-4,6-dichloro-*s*-triazine) with OH groups of cellulose

radiation and plasma treatment in presence of oxygen or air increase the amount of aldehyde groups and change the surface energy of the wood fiber. Cold plasma treatment of wood fibers in presence of oxygen also introduces surface cross linkings by production of reactive free radicals and groups. But the chief aim of all these physical treatments is to improve the adhesion between the polymer and the WPC surface. Before we study in detail about the surface modification of wood fibers a brief review of the adhesion theory is presented below.

### *2.5.1 Forces Involved in Adhesion*

The state of adhesion between two substrates is sustained by the interfacial forces that comprise of interlocking action or valence force action or in some cases both. Interfacial forces also known as adhesive forces originate from the forces between atoms or molecules of different substrates. If the adhesive force is weak, adhesion failure occurs at this interface between the wood fiber and polymer matrix. This force can be broadly divided into two groups: the primary forces and secondary forces. Primary force also known as short range force is arising from chemical bonding. Chemical bonding compromises of interlinking between molecules of the substrate by covalent, ionic or metallic bonds. The second forces also called as long range forces originate from physical attraction or due to the formation of weak Van der Waal's bonds like hydrogen bonds between the two substrates that are in intimate contact. Generally, the interactions between the substrates involve both of the above forces.

## 2.5.2 *Mechanism of Adhesion With Respect to WPC*

There is no single theory of adhesion which can satisfactorily explain all the interactions between the adhesive and substrate. Several existing theories provide various perspectives, each theory is applicable to certain applications but there is no universal mechanism of adhesion available as yet. Of the many adhesion mechanisms found in literature mechanical interlocking, adsorption and diffusion theory, boundary layer can be applied to WPC. A brief discussion of these theories as applied to WPC is given below

### 2.5.2.1 Mechanical Interlocking

The surface of wood powder (both compacted and powder form) and many wood fibers are not perfectly smooth and contain crevices, peaks and valleys. If the viscosity of the polymer is not high, the polymer melt can flow and fill these micro-cavities of the porous wood powder. When the adhesive solidifies the substrates are held together by mechanical anchoring. This mechanism is often used in polymer composites by etching the polymer surface to increase the surface roughness thereby increasing the contact area for adhesive penetration and mechanical interlocking of the substrate.

### 2.5.2.2 Adsorption and Diffusion Theory

Adsorption and diffusion theory states the involvement of two stages in polymer–substrate adhesion mechanism. In the first stage, the adsorption stage the two materials, polymer and wood powder should have intimate contact which is in turn governed by two actions: spreading and penetration. Once good wetting occurs permanent adhesion is developed through molecular attractions. The molecular attractions are developed due to any one of these: covalent, electrostatic, metallic and van der Waal's. Good wetting between the substrates can lead to inter-diffusion of molecules between the wood powder and polymer matrix. The degree and extent of diffusion depends mainly on the chemical compatibility of the two materials and penetrability of the substrate.

### 2.5.2.3 Boundary Layer Theory

In this theory, it is proposed that treatment of fillers with coupling agent or compatibilizers leads to formation of a layer that adheres to the surface of the filler. This layer can be strong as in deformable layers (wherein a coupling agent produces a tough and flexible layer) or a restrained layer (wherein coupling agents develop highly crosslinked surface region with a modulus intermediate between the substrate and bulk) or formation of a weak boundary layer wherein the coupling agent simple wets the surface of the filler.

The development of a definitive theory for the mechanism of bonding by coupling agents in composites is a complex problem. The main chemical bonding theory alone is not sufficient. So, the consideration of other concepts appears to be necessary, which include the morphology of the interface, the acid–base reactions at the interface, surface energy and the wetting phenomena needs to be considered. In practice a combination of all the three layers in coupling agent treated wood fiber can be expected since the surface of wood fiber is chemically heterogeneous.

### 2.5.3 Plasma and Corona Treatment

Plasma and corona treatment of wood and natural fibers has been recently explored. These methods are physical and do not involve production of considerable amounts of hazardous waste by products as with the traditional chemical treatments like mercerization, acetylation, grafting and coupling agents. Especially plasma treatment is very environmentally friendly and by utilizing plasma treatment one can also impart additional functional groups like amine, sulfonates, etc., onto the wood surface. Plasma is ionized gas containing a mixture of ions, electrons, neutral and excited molecules and photons. Many new techniques like atmospheric air pressure plasma (AARP) are also being used to modify the surface of wood fibers. Depending on the nature and composition of feed gases, variety of surface modifications which can increase or decrease the surface energy, grafting of reactive free groups and even surface cross linking can be achieved. Plasma treatment of wood fibers not only removes the surface impurities but also defibrillate the fibers making the natural fiber porous thereby increasing the probability of mechanical interlocking of polymer chains on the surface of wood fibers. Another advantage of plasma treatment is the time consumed. Wood fibers are normally treated with plasma for 1–3 min which is far shorter than the time taken by traditional methods which can run for days (Fig. 2.27).

Figure 2.28 shows the morphology of plasma treated wood fiber. Untreated wood fiber as shown in Fig. 2.28a has a smooth surface since the primary walls of wood fibers consist of fats, fatty acids, fatty alcohols, phenols, terpenes, steroids and waxes. This outer layer covers a secondary layer which is a complex of hemi and hollow cellulose, lignin, pectins and proteins. Plasma treatment not only effectively penetrates both these layers but if the time of plasma treatment is high it can destroy the outer layer to a considerable extent. Figure 2.28b shows a very rough surface consisting of large amounts of defects and cracks, pits, corrugations, etc. Plasma treatment is nothing but physical etching that causes the already existing flaws and surface instabilities of fiber surface to expand. Increasing the plasma treatment time to beyond 3 min is not advisable in order to avoid total

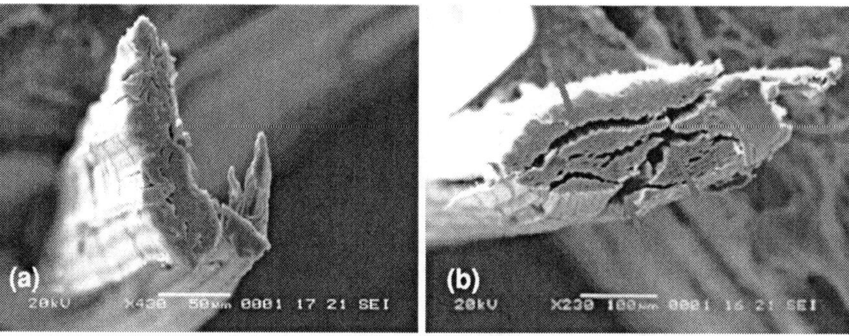


**Fig. 2.27** Cross section of wood fiber **a** before and **b** after plasma treatment. *Arrows* indicate the presence of cracks

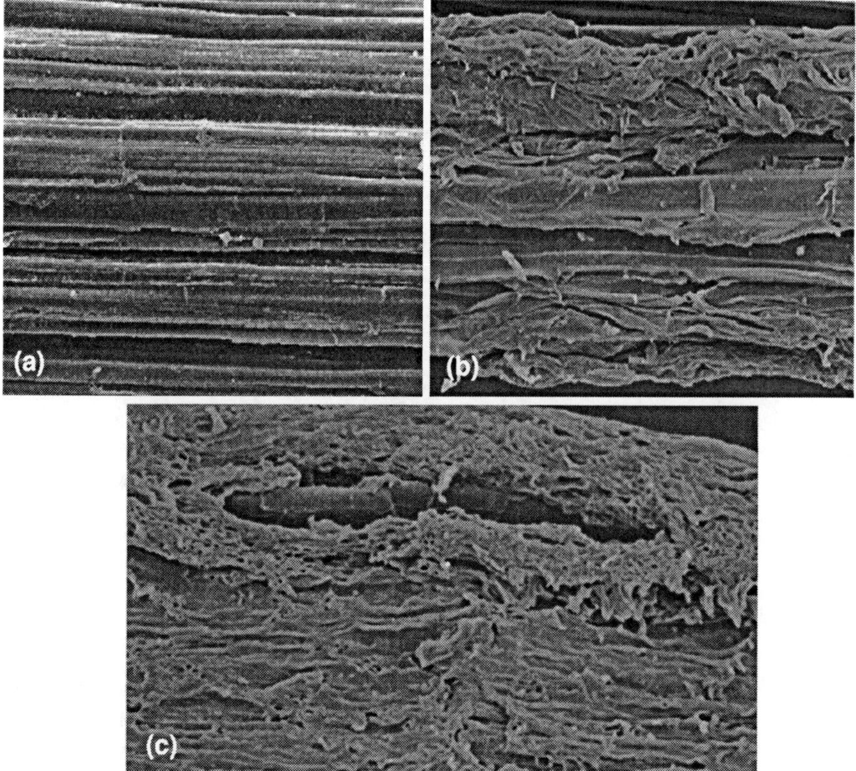


**Fig. 2.28** Surface morphology of wood fiber **a** before and **b** and **c** after plasma treatment

deterioration of the wood fibers. Increased plasma exposure of wood fibers result in "mechanically weak boundary layers" due to mechanical fiber degradation ex: heavily roughened surfaces and loosely attached fibrils. Another disadvantage of plasma treatment is the migration of hydrophilic extract.

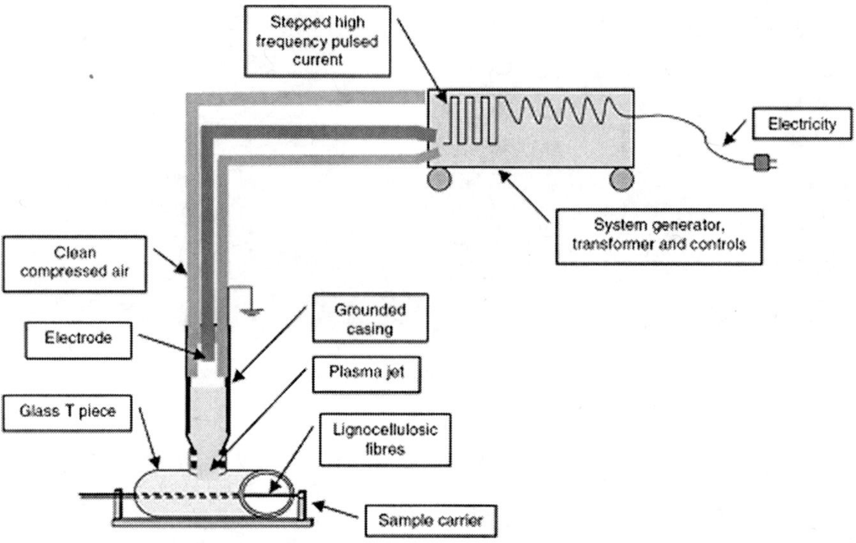


**Fig. 2.29** Schematic representation of plasma treatment method

Typical plasma treatment equipment is shown in Fig. 2.29.

Figure 2.29 shows an open air plasma system. The maximum power can range from 1 to 50 kW for variable duration of time. This equipment can be continuous when fitted with a conveyor belt mounted sample holder. The plasma gun is an inverted glass T-piece, wherein the middle inlet was positioned under the plasma nozzle in an attempt to confine the active plasma species over the fibers for the duration of treatment. The height between the sample and plasma gun can be adjusted and sufficient care must be taken to avoid thermal degradation of the wood fibers. Excessive time of plasma treatment and very short distance between the wood fibers and plasma gun can cause thermally induced changes which can be visually observed like changes in color of fibers, carbonization, etc.

### 2.5.4 *Ozone Treatment*

Ozone treatment is a method that has been proven to oxidize a wide variety of surfaces and has been proven to be very effective in enhancing adhesion between various substrates.

## 2.6 Chemical Treatments

Because of the low interfacial properties between fiber and polymer matrix which reduces the potential of natural fibers as reinforcing agents due to the hydrophilic nature of natural fibers chemical modifications are considered to optimize the interface of fibers. Chemical treatment of fibers will activate hydroxyl groups or introduce new moieties that can effectively interlock with the matrix. Most chemical agents used to modify wood fibers possess two functions. The first function is to react with the hydroxyl groups present on the surface of fibers and second is to react with the functional groups of the polymer matrix. The general mechanisms of coupling agents can be outlined by the following mechanisms.

(a) Elimination of weak boundary layers.
(b) Production of a tough and flexible layer.
(c) Development of highly cross-linked interface region.
(d) Improvement of wetting between the polymer and fiber.
(e) Possible formation of chemical interactions either by covalent bonds or hydrogen bonds between the fiber and the polymer.
(f) Alteration of acidity or surface energy of the fibers.

Before treating wood fibers with coupling agents, wood is treated with alkali or ammonia to A brief review of various chemical treatments used in WPC is given below.

### 2.6.1 Steam Explosion

Steam explosion of wood flour and wood fibers has been long known for preparing plastic like molded wood products (ideally with a small quantity of resin binder added to it). Steam exploded wood is an excellent additive to WPC because wood components are decomposed into fragments of low molecular weight, besides leaching out substantial portions of lignin, wax and oily impurities. Steam explosion of wood is also known to decrease hydrophilicity by dehydration. Steam exploded wood is very advantageous especially from rheological point of view. A typical steam explosion set up is shown in the Fig. 2.30.

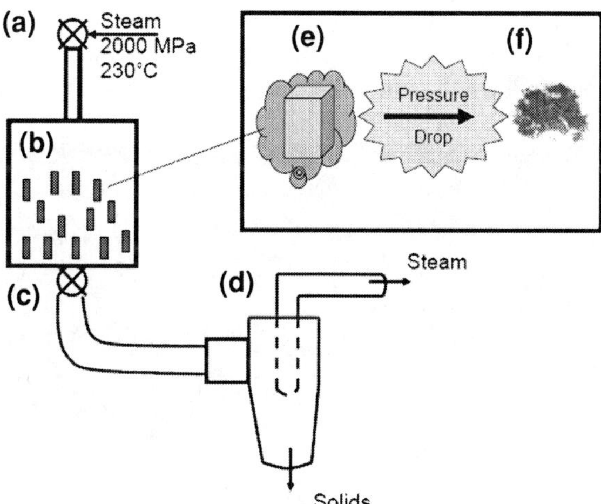


**Fig. 2.30** Flow charts for batch steam of wood or natural fibers

Steam explosion technology as a technique to defibrillate lignocellulosic materials like wood and natural fibers has been in use since 60 years with both batch and continuous (stake) reactors available in the market. In steam explosion, wood particles are first treated with high pressure steam for short periods of time, followed by sudden decompression (explosion) which results in fiberization or "mulching" by a complex physico-chemical reaction. The parameters controlling the steam explosion technique are reaction temperature ($T_r$) and retention time ($t$). The degree of mulching is generally expressed by the term "severity" and is given by the expression:

$$R_o = \int_0^t \exp[(T_r - T_b)/14.75]dt \tag{2.1}$$

where $T_b$ is the base temperature (temperature of steam inlet, normally 100°C). Chemically speaking steam explosion is an auto-hydrolysis process and has these following effects.

- Cleavage of some accessible glycosidic links.
- Cleavage of $\beta$-ether linkages of lignin.
- Cleavage of lignin-carbohydrate complex bonds.
- Minor chemical modification of lignin and carbohydrates.

The effect of steam explosion in soft woods like pine is much more dramatic than hard and seasoned woods. Figure 2.31 illustrates the changes in morphology of steam exploded pine wood as studied by SEM. Extensive defibrillation of fiber bundles in steam exploded sample can be observed.

A big advantage of steam explosion is that one can recover the lignins, cellulose and hemicellulose from the exploded fibers either by simple fractionation or by treating it with alkalis to form alkali soluble and alkali-insoluble fractions. Steam explosion increases the porosity of the wood fiber thereby increasing the chances of mechanical interlocking. Figure 2.32 shows the morphology of untreated and steam exploded wood fiber in polyethylene matrix.

The conditions of steam explosion and wood species influence the resulting morphology. Steam exploded wood and other biomass is widely used to form

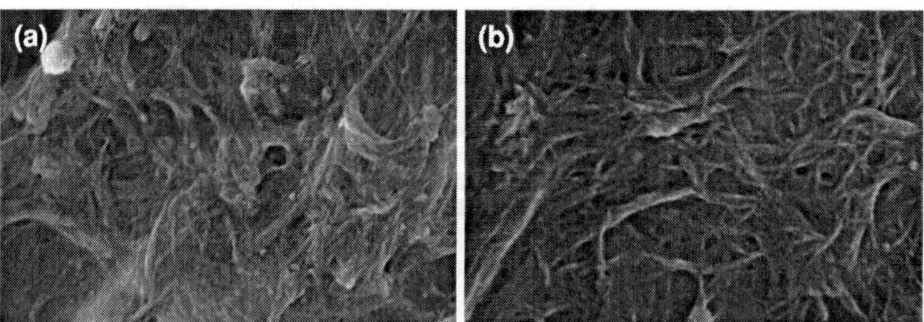


**Fig. 2.31** SEM microphotographs of **a** untreated and **b** steam exploded pine wood

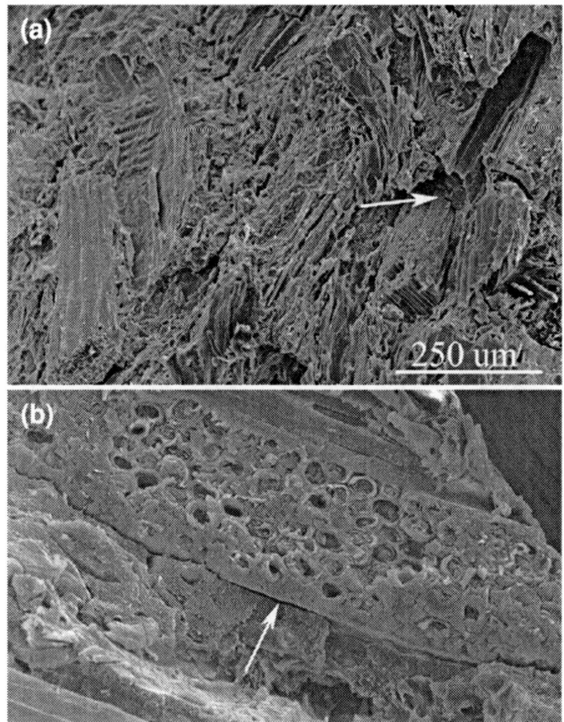


**Fig. 2.32** Morphology of **a** untreated and **b** steam exploded wood fibers in polyethylene

'binderless' panels, but this requires very high temperatures. Elevated temperatures above the softening temperature of lignin are required for composites to achieve desired physical and mechanical properties. Before steam explosion, wood fibers are treated with simple pretreatment with dilute sulfuric acid is also used. However additional lignin should be added in this type of acid treated wood.

## *2.6.2 Alkali Treatment*

Alkaline treatment or mercerization is one of the oldest, cost effective and most used chemical treatments of natural fibers to be used in thermoplastics and thermosets. Mercerization of cotton fibers with alkaline solutions has been known to humanity for centuries and was widely practiced in Egyptian and Indian civilization from prehistoric days. The reaction of NaOH with cellulose present in wood is shown as:

$$\text{Cell - OH} + \text{NaOH} \rightarrow \text{Cell - O} - \text{Na+} + H_2O + \text{surface impurities}$$

Alkali treatment is not only a highly efficient and effective way of removing surface impurities but also it can remove majority of lignins, wax and oils covering the external surface of the fiber cell wall. Alkali treatment also depolymerizes the

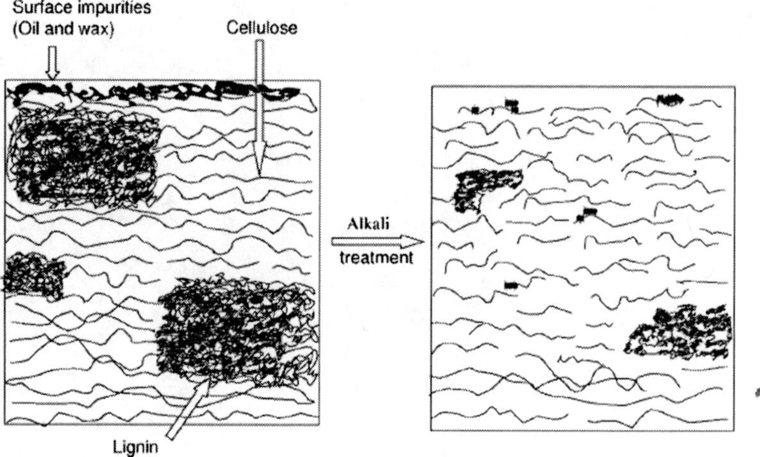


**Fig. 2.33** Schematic representation of raw and alkali treated wood and natural fibers

native cellulose structure and exposes short length crystallites as shown in Fig. 2.33

Alkali treatment also removes hemicellulose from surface of wood fibers making the interfibrillar region less dense and less rigid making the fibrils more likely to rearrange themselves along the direction of tensile deformation. Another important aspect of alkali treatment is the increase in the percentage of crystallinity commonly referred as crystallinity index. XRD studies also showed that by alkali treatment there is a decrease in spiral angle and increase in molecular orientation. Figure 2.34 shows the XRD diffractographs of alkali and steam treated natural fibers wherein an increase in crystallinity can be clearly observed.

The changes in crystallinity with alkali treatment can be observed from the changes in the [200] lattice plane of cellulose. Cellulose of wood has a order–disorder kind of structure where in hard crystalline segments of cellulose are weakly linked together by soft amorphous chains. The extent of this amorphous–crystalline structure depends on many factors like type of species, tree location, aging, seasoning, etc. Alkali treatment changes these intermediate regions with higher concentrations of NaOH making this region disordered thereby increasing the crystallinity of the wood fiber. Alkali treatment also increases the orientation distribution of cellulose crystallites in the fiber samples. Figure 2.35 shows plot of alkali and steam treatments on the crystallite orientation in pineapple, ramie and sansevieria natural fibers.

Alkali and steam treatments also change the texture of natural fibers by making the surface more rougher thereby increasing the probability of mechanical interlocking when these treated fibers are added into the composite. Figure 2.36 shows the morphology of alkali and steam treated sansevieria fibers. Alkali treated fibers appear to be cleaner and fiber bundles were more separated with serrated structure.

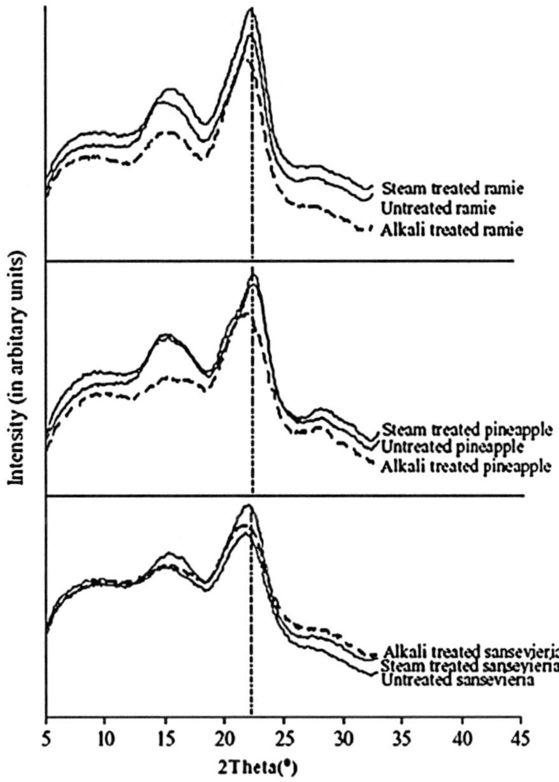


**Fig. 2.34** X-ray diffractographs of untreated and treated natural fibers

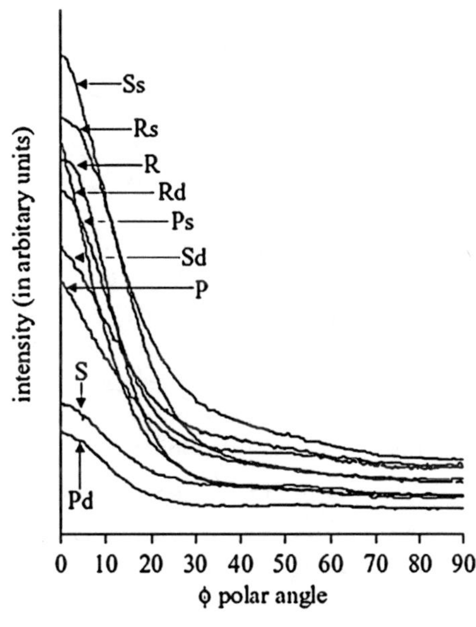


**Fig. 2.35** Crystallites orientation factors in natural fibers: P, S and R denote pineapple, sansevieria and Ramie fibers, subscript *s* and *d* refers to NaOH and steam treatments, respectively

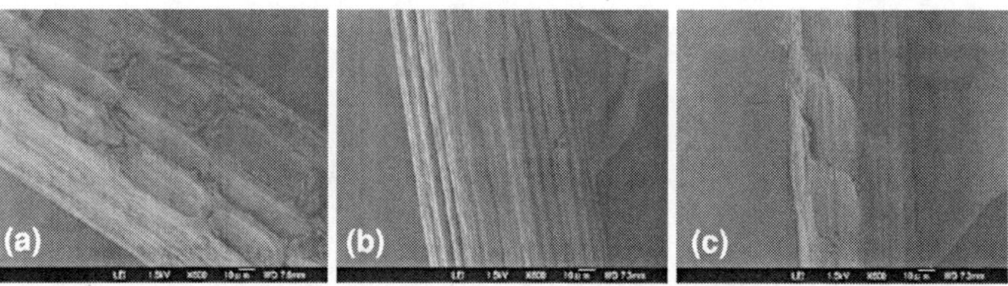


**Fig. 2.36** SEM micrographs of untreated (**a**), alkali (**b**) and steam (**c**) treated sansevieria fibers

SEM micrographs show that treating of natural fibers with alkalis changes the porous bundle structure.

However it must be mentioned that alkali treatment of natural fibers causes decrease in tensile strength and Young's modulus of fibers due to decrease in degree of crystallinity and extent of crystallite orientation. Higher concentration of NaOH solution and higher treatment time also causes substantial reduction in tensile properties of natural fibers. Besides higher concentration of NaOH also lead to extensive damage of fibers leading to drastic reduction and difficulties in handling of the fibers. Figure 2.37 shows the extent of damage in jute fibers at various treatments of NaOH.

So the choice of optimum concentration and treatment time is a prerequisite. Generally the rule of thumb in industry is 5–9% NaOH solutions for 30–45 min will lead to optimal properties without any drastic change in the properties of fibers.

Besides NaOH, other forms of chemical treatments can be used to remove lignins and oily and waxy substances from wood powder. These treatments can be divided into two groups:

Group 1: Multiple step slightly acidic solutions like

- Acidified sodium chlorite method which has ($CH_3COOH + NaClO_2$) for 12 h at 60°C, followed by steeping in 1 N NaOH for 1 h at 60°C and 0.5% HCl for 1 h at 15–20°C.
- Modified sodium chlorite method with Amox treatment which consists of treating wood fibers with ($CH_3COOH + NaClO_2$) for 12 h at 60°C, followed by steeping in 1 N (NH4)2$C_2O_4$(AMox) for 1 h at 60°C and finally with 1% NaOH for 1 h at 15–20°C.
- Peracetic acid method: steeping of wood fibers in equal 1:10 liquor composed of equal quantities of glacial acetic acid and 20% hydrogen peroxide.
  Group 2: Based on pulping process of paper

- Soda pulping: 17% NaOH for 1.5 h at 70°C
- Kraft pulping: 17% NaOH and 25% sulfidity for 1.5 h at 70°C
- Kraft pulping for soft woods (like Pine): 20% NaOH and 25% sulfidity for 1.5 h at 70°C.

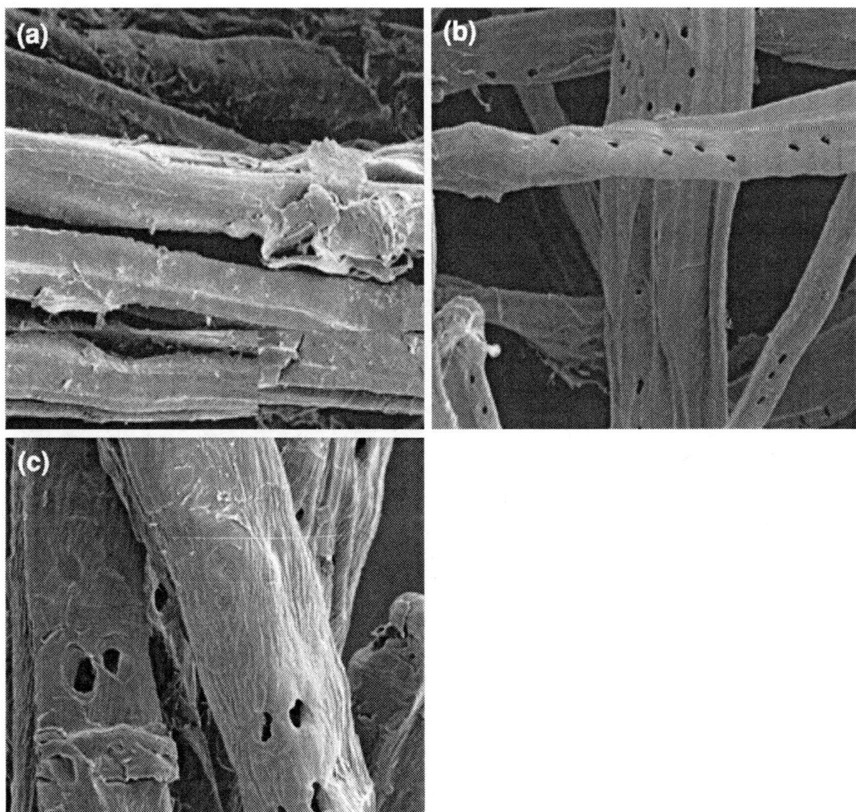


**Fig. 2.37** Damage caused to jute fibers by extensive NaOH treatment

### *2.6.3 Srearic Acid and Wax Treatment*

Another type of surface treatment of wood fibers is by stearic acid. Stearic acid treatment of wood fiber imparts better dispersion primarily by reducing the surface tension of the wood fibers thereby reducing the polymer–wood fiber contact tensions. However, this dispersing effectiveness of stearic acid depends on the method of incorporation. The most effective way is to treat wood fibers from a solution phase of stearic acid. However, simultaneously with positive influence on dispersion, incorporation of stearic acid leads to considerable decrease of melt flow index (MFI) of the composite. This can be attributed to the interactions of the acid groups of stearic acid with hydroxyl groups of wood fiber which results in decreasing of the mobility of polymer chains at the interface. Paraffin wax treatment of wood fibers has also been investigated. The effect of paraffin wax treatment of wood fibers is analogous to stearic acid but the decrease in MFI is not as severe in stearic acid treated wood powders. Besides paraffin wax treatment also decreases the moisture uptake of the composites to a considerable extent since

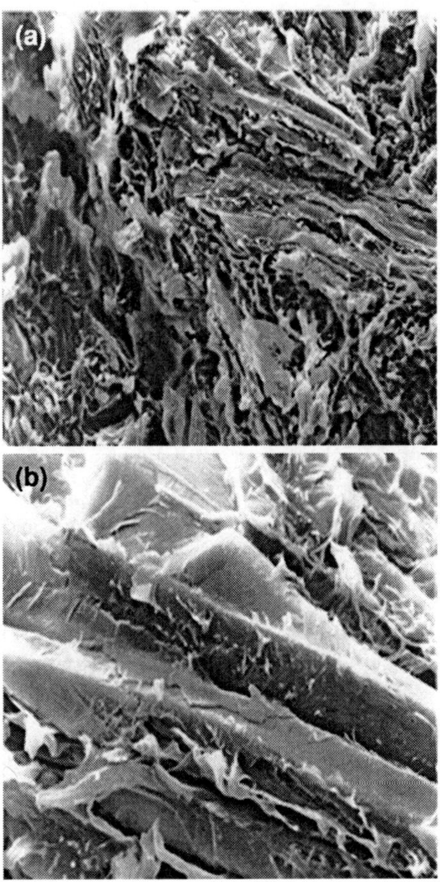


**Fig. 2.38** Morphology of **a** untreated and **b** paraffin wax treated wood fibers

treatment of paraffin wax imparts hydrophobicity of the fibers. Besides this low viscosity and high processing temperatures during processing also may melt the wax and it can possibly penetrate into WF pores. However both stearic acid and paraffin wax treatment of wood powder is known to cause considerable decrease in the hydroxyl groups of the wood fiber thereby decreasing the polymer–wood fiber interfacial contacts thereby having an detrimental effect on performance characteristics of the composites. Figure 2.38 shows SEM micrographs of untreated and wax treated wood fibers.

## 2.7 Some Novel (Enzymatic) Treatments

Besides these chemical and physical ways of treatments, some novel treatments of wood fibers are also being recently explored. The most interesting of these are enzymatic treatment and by bacteria.

The use of enzymatic treatment to enhance properties of WPC is less common but offers a unique way of modifying the chemical structure and ultra-structure of the fiber. Enzymes are biological catalysts that increase the rate of chemical reactions and the reactants enzymes are called as substrates. There are literally millions of enzymes bestowed upon humanity by Mother Nature, all of which are proteins. However, an enzyme is always accompanied by a non-protein component called a co-factor without which the enzyme lacks any activity. Enzymes exhibit remarkable group specificity in that they may act on several different though closely related substrates to catalyze a reaction involving a particular chemical group. One more advantage of enzymatic treatment of wood fibers is the non-necessity of previous pre-treatments except for size reduction. This makes enzyme treatment of wood fibers very attractive environmentally. For an effective enzyme action, three different points of interactions between the enzymes and substrates are essential. These can be classified broadly as binding, catalytic and active sites. The binding and catalytic sites should invariably be either polar or non-polar

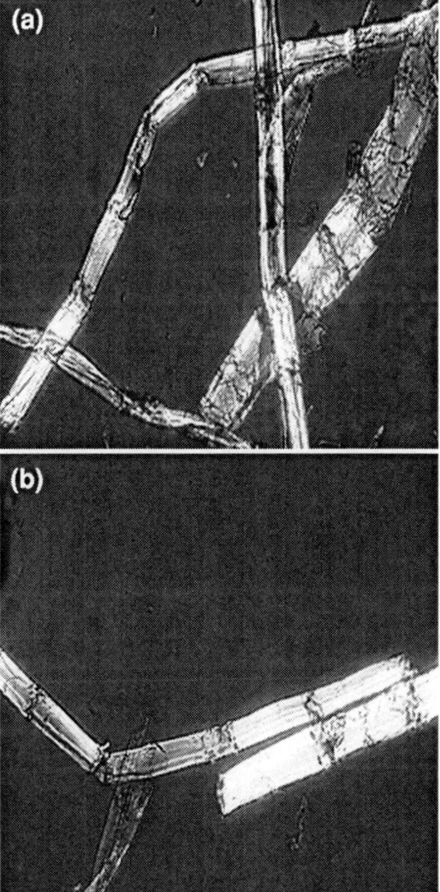


**Fig. 2.39** Enzyme treated fibers for 1 h **a** 1%, **b** 5% enzyme solution

amino acid residues, thus creating an arrangement of hydrophilic and hydrophobic environment. In wood fibers, it is commonly known as cellulose binding domains (CBDs). These CBDs can be further sub divided into several different families based on amino acid sequence similarities.

Now the question arises how enzymes can modify wood fibers. The hypothesis underlying this is the superb selectivity of enzymes to degrade specific components of the fiber and softening of the bulk accompanied by increase in fiber structure. Enzymatic action of enzymes on wood fibers enhances local defects, thereby providing localized 'hinge points' in the fibers. Enzyme treatment also can effectively hydrolyze the hemicellulose of wood fibers. Hemicellulose is the main adhesive that holds the fibers together. Enzyme treatment also reduced degree of polymerization level by effectively removing the xylan content. One more advantage of enzyme treatment is it superb action to smoothen the fibers. Enzymes (for some unknown reason) seem to show much localized attacks mostly in kinks and nodes of the fiber walls thereby effectively smoothening the fiber surface. This however may result in degree of mechanical interlocking between the fibers and polymer matrix.

Excessive degradation is a serious disadvantage when one uses enzymes to treat wood fibers. This problem is more exaggerated if the wood fibers are a mixture of fibers from different species, different ages, regions, etc. Excessive degradation occurs when enzymes aggressively attack the outer most layers of the fiber walls

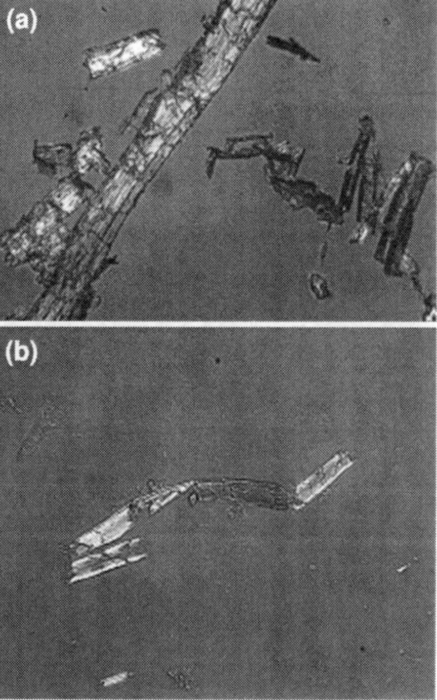


**Fig. 2.40** Enzyme treated fibers for 23 h **a** 1%, **b** 5% enzyme solution

ultimately reducing the thickness of cell walls. This is particularly serious if one uses cellulose enzymes and trichoderma celluloses. Besides the type of species of enzymes, the concentration and time of treatment affects the properties of fibers (Figs. 2.39, 2.40).

Though Proctor & Gamble patented and commercially marketed enzyme treatment to obtain extra-soft tissue papers from bagasse pulp way back in 1990s, its utility in WPC is considerably low. This is because enzyme treatment is very expensive and time consuming. Enzyme solutions such as Pectinex Ultra SP-L, a cellulase enzyme are commercially available.

## References


1. Klason, C., Kubát, J., Strömvall, H.-E.: "The efficiency of cellulosic fillers in common thermoplastics, part I. Filling without processing aids or coupling agents". Int. J. Polym. Mater. **10**, 159–187 (1984)
2. Maldas, D., Kokta, B.V.: Surface modification of wood fibers using maleic anhydride and isocyanate as coating components and their performance in polystyrene composites. J. Adhesion Sci. Technol. **5**(9), 727–740 (1991)
3. Mankowski, M., Morrell, J.J.: Patterns of fungal attack in wood–plastic composites following exposure in a soil block test. Wood Fiber Sci. **32**(3), 340–345 (2000)
4. Verhey, S.A., Laks, P.E., Richter, D.L.: Laboratory decay resistance of woodfiber/thermoplastic composites. For. Prod. J. **51**(9), 44–49
5. Youngquist, J.A.: Unlikely partners? The marriage of wood and nonwood materials. For. Prod. J. **45**(10), 25–30 (1995)
6. Gaylord, N.G.: Compatibilization of hydroxyl containing materials and thermoplastic polymers. United States Patent Office no 3,645,939 (1972)
7. Coran, A.Y., Patel, R.: US Patent 4,323,625 (1982)
8. Geottler, L.A.: US Patent 4,376,144 (1983)
9. Meyer, J.A.: Crosslinking affects sanding properties of wood–plastic. For. Prod. J. **18**(5), 89 (1968)
10. Nakamura, T., Okamura, M., Moriguchi, Y., Hayase, T.: US Patent 4,404,437 (1983)
11. Woodhams, R.T., Thomas, G., Rodgers, D.K.: Wood fibers as reinforcing fillers for polyolefins. Polym. Eng. Sci. **24**(15), 1166–1171 (1984)
12. Xanthos, M.: Processing conditions and coupling agent effects in polypropylene/wood flour composites. Plast. Rubber Process. Appl. **3**(3), 223–228 (1983)
13. Schneider, M.H., Brebner, K.I.: Wood–polymer combinations: the chemical modification of wood by alkoxysilane coupling agents. Wood Sci. Technol. **19**(1), 67–73 (1985)

# Chapter 3
# Process and Machinery Used for WPC

## 3.1 The Manufacturing Process

The feature common to all polymeric composite processes is the combining of a resin, a curing agent, some type of reinforcing fiber, and in some cases a solvent. Typically, heat and pressure are used to shape and "cure" the mixture into a finished part. In composites, the resin acts to hold the fibers together and protect them, and to transfer the load to the fibers in the fabricated composite part. The curing agent, also known as hardener, acts as a catalyst and helps in curing the resin to a hard plastic. The reinforcing fiber imparts strength and other required properties to the composite.

## 3.2 Polymer Matrix Composites

Advanced composites exhibit desirable physical and chemical properties that include light weight coupled with high stiffness and strength along the direction of the reinforcing fiber, dimensional stability, temperature and chemical resistance, flex performance, and relatively easy processing. Advanced composites are replacing metal components in many uses, particularly in the aerospace industry.

### *3.2.1 Resins*

The resin systems used to manufacture advanced composites are of two basic types: thermosetting and thermoplastic. Thermosetting resins predominate today, while thermoplastics have only a minor role in advanced composites manufacture.

J. K. Kim and K. Pal, *Recent Advances in the Processing of Wood–Plastic Composites*, Engineering Materials, 32, DOI: 10.1007/978-3-642-14877-4_3,

### 3.2.2 Thermosets

Thermoset resins require addition of a curing agent or hardener and impregnation onto a reinforcing material, followed by a curing step to produce a cured or finished part. Once cured, the part cannot be changed or reformed, except for finishing. Some of the more common thermosets include:

- epoxies
- polyurethanes
- phenolic and amino resins
- bismaleimides (BMI, polyimides)
- polyamides.

Of these, epoxies are the most commonly used in today's PMC industry. Epoxy resins have been in use in the industry for over 40 years. The basic epoxy compounds most commonly used in industry are the reaction product of Epichlorohydrin and Bisphenol-A. Epoxy compounds are also referred to as glycidyl compounds. There are several types of epoxy compounds including glycidyl ethers (or diglycidyl ethers), glycidyl esters, and glycidyl amines. Several of these compounds are reactive diluents and are sometimes added to the basic resin to modify performance characteristics. The epoxy molecule can also be expanded or cross-linked with other molecules to form a wide variety of resin products, each with distinct performance characteristics. These resins range from low-viscosity liquids to high-molecular weight solids. Typically they are high-viscosity liquids.

The second of the essential ingredients of an advanced composite system is the curing agent or hardener. These compounds are very important because they control the reaction rate and determine the performance characteristics of the finished part. Since these compounds act as catalysts for the reaction, they must contain active sites on their molecules.

Some of the most commonly used curing agents in the advanced composite industry are the aromatic amines. Two of the most common are 4,4′-methylene-dianiline (MDA) and 4,4′-sulfonyldianiline (DDS). Less hazardous curing agents have been introduced into the industry as MDA has been phased out.

Several other types of curing agents are also used in the advanced composite industry. These include aliphatic and cycloaliphatic amines, polyaminoamides, amides, and anhydrides. Again, the choice of curing agent depends on the cure and performance characteristics desired for the finished part.

Polyurethanes are another group of resins used in advanced composite processes. These compounds are formed by reacting the polyol component with an isocyanate compound, typically toluene di-isocyanate (TDI); methylene di-isocyanate (MDI) and hexamethylene di-isocyanate (HDI) are also widely used.

Phenolic and amino resins are another group of PMC resins. They are used extensively in aircraft interiors because of their exceptional low smoke and heat release properties in the event of a fire.

The bismaleimides and polyamides are relative newcomers to the advanced composite industry and are used for high temperature applications.

### 3.2.3 *Thermoplastics*

Thermoplastics currently represent a relatively small part of the PMC industry. They are typically supplied as nonreactive solids (no chemical reaction occurs during processing) and require only heat and pressure to form the finished part. Unlike the thermosets, the thermoplastics can usually be reheated and reformed into another shape, if desired.

### 3.2.4 *Reinforcements*

Fiber reinforcement materials are added to the resin system to provide strength to the finished part. The selection of reinforcement material is based on the properties desired in the finished product. These materials do not react with the resin but are an integral part of the advanced composite system.

Three basic types of fiber reinforcement materials in use in the advanced composite industry are:

- carbon/graphite
- aramid
- glass fibers.

Fibers used in advanced composite manufacture come in various forms, including:

- yarns
- rovings
- chopped strands
- woven fabric
- mats

Each of these has its own special application. When prepreg materials are used in parts manufacture, woven fabric or mats are required. In processes such as filament wet winding or pultrusion, yarns and rovings are used.

The most commonly used reinforcement materials are carbon/graphite fibers (the terms graphite and carbon are often used interchangeably). This is due to the fact that many of the desired performance characteristics require the use of carbon/graphite fibers. Currently, these fibers are produced from three types of materials known as precursor fibers:

- polyacrylonitrile (PAN)
- rayon
- petroleum pitch.

The carbon/graphite fibers are produced by the controlled burning off of the oxygen, nitrogen, and other noncarbon parts of the precursor fiber, leaving only carbon in the fiber. Following this burning off (or oxidizing) step, the fibers are run through a furnace to produce either carbon or graphite fibers. Carbon fibers are produced at furnace temperatures of 1,000–2,000°C, while graphite fibers require temperatures of 2,000–3,000°C. At these temperatures the carbon atoms in the fibers are rearranged to impart the required characteristics to the finished fiber. The PAN-based fiber is the more commonly used precursor in the advanced composite industry today.

Aramid fibers are another human-made product. These fibers are produced by manufacturing the basic polymer, then spinning it into either a paper-like configuration or into fiber. Aramid fibers have several useful characteristics:

- high strength and modulus
- temperature stability
- flex performance
- dimensional stability
- chemical resistance
- textile processibility.

Textile (continuous filament) glass fibers are the type used in composite reinforcement. These fibers differ from the wool type in that they are die-drawn rather than spun.

A number of solvents are used in the advanced composites industry. These may be introduced into the workplace in three basic ways:

- as part of the resin or curing agent
- during the manufacturing process
- as part of the cleanup process.

## 3.3 Major Processes Involved for Composite Preparation

The major processes used in the advanced composites industry are discussed below. The processes vary widely in type of equipment used. Several of the processes are automated; however, some are manual and require worker contact with the part during manufacture. The basic process types are described below.

*Formulation* is the process where the resin, curing agent, and any other component required are mixed together. This process may involve adding the components manually into a small mixing vessel or, in the case of larger processes, the components may be pumped into a mixing vessel.

*Prepregging* is the process where the resin and curing agent mixture are impregnated into the reinforcing fiber. These impregnated reinforcements (also known as prepregs) take three main forms: woven fabrics, roving, and unidirectional tape. Fabrics and tapes are provided as continuous rolls in widths up to 72 in and lengths up to several hundred feet. The fabric or tape thickness constitutes one ply in the construction of a multi-ply lay-up. Impregnated roving is wound onto cores or bobbins and is used for filament winding. Once the resin mixture has been impregnated onto the fibers, the prepreg must be stored in a refrigerator or freezer until ready for use in the manufacturing process. This cold storage prevents the chemical reaction from occurring prematurely. Prepreg materials are used widely in the advanced composite industry, particularly in aircraft and aerospace.

*Open Molding* processes are those where the part being manufactured is exposed to the atmosphere. The worker typically handles the part manually, and there is a higher potential for exposure. The resin mixture may be a liquid being formed onto a reinforcing material or it may be in the form of a prepreg material being formed for final cure.

*Closed Molding* processes are those in which all or part of the manufacture takes place in a closed vessel or chamber. The liquid resin mixture or prepreg material may be handled or formed manually into the container for the curing step. In the case of liquid resin mixtures, these may be pumped into the container, usually a mold of some type, for the curing step.

*Sequential or batch* processes involve manufacture of a single part at a time, in sequence. This type of process is usually required where the part being made is small and complex in shape, when the curing phase is critical, when finishing work must be minimized, or where a small number of parts is involved.

*Continuous* processes are typically automated to some degree and are used to produce larger numbers of identical parts relatively quickly. These processes are typified by pumping of the resin mixture into the mold, followed by closed curing.

## 3.4 Description of Shaping Processes

A brief description of each process as follows

### *3.4.1 Resin Formulation*

Resin formulation consists of mixing epoxy or other resins with other ingredients to achieve desired performance parameters. These ingredients may be curing agents, accelerators, reactive diluents, pigments, etc.

### *3.4.2 Prepregging*

Prepregging involves the application of formulated resin products, in solution or molten form, to reinforcement such as carbon, fiberglass or aramid fiber or cloth. The reinforcement is saturated by dipping through the liquid resin. In an alternate

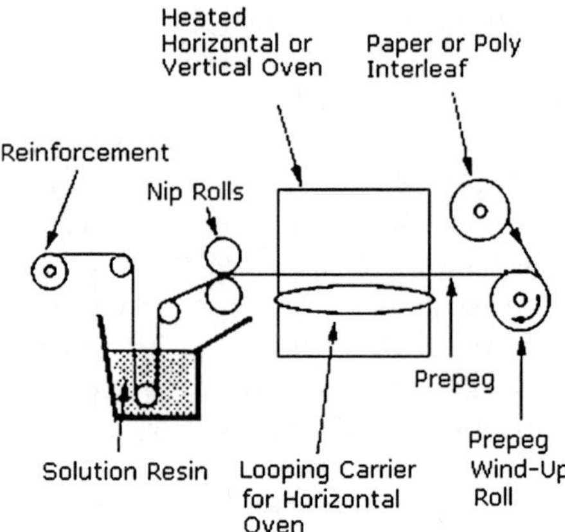


**Fig. 3.1** Schematic diagram of prepregging

method called a Hot Melt Process the resin is impregnated through heat and pressure. The Hot Melt System uses resins with a very low percentage of solvents Fig. 3.1.

### *3.4.3 Wet Filament Winding*

In the filament wet winding process, continuous fiber reinforcement materials are drawn through a container of resin mixture and formed onto a rotating mandrel to achieve the desired shape. After winding, the part is cured in an oven. This process can also used preimpregnated fiber tows called towpregs (Fig. 3.2).

### *3.4.4 Hand Lay-Up of Prepreg*

The prepreg product is trimmed and laid down over a mold where it is formed to the desired shape. Several layers may be required. After forming, a vacuum bag is

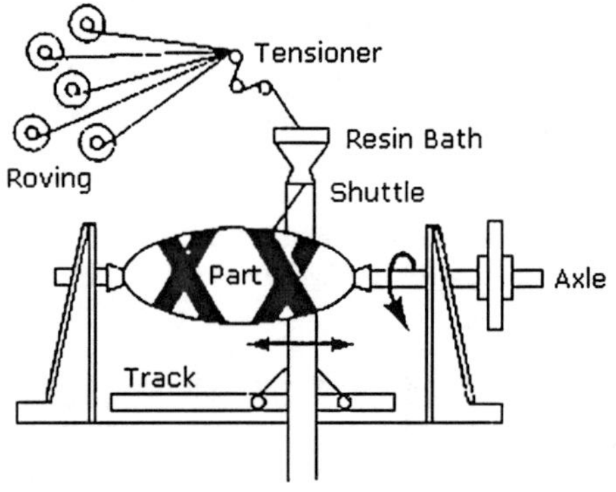


**Fig. 3.2** Wet filament winding

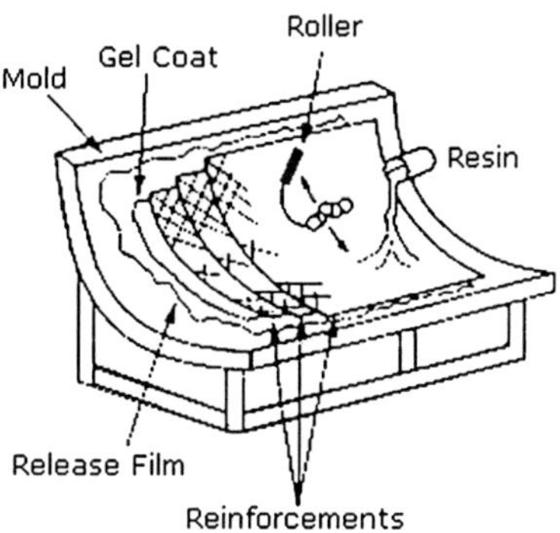


**Fig. 3.3** Diagram of hand lay-up

sealed around the lay-up. Vacuum is pulled on the raw prepreg to remove air, compact the part and serve as a barrier when the assembly is placed in an autoclave for cure under heat and pressure. Oven cures (under vacuum only) may be used for non-structural parts (Fig. 3.3).

### *3.4.5 Automated Tape Placement*

In this process, the prepreg tape material is fed through an automated tape application machine (robot). The tape is applied across the surface of a mold in multiple layers by the preprogrammed robot.

### *3.4.6 Resin Transfer Molding*

Resin transfer molding is used when parts with two smooth surfaces are required or when a low-pressure molding process is advantageous. Fiber reinforcement fabric or mat is laid by hand into a mold and resin mixture is poured or injected into the mold cavity. The part is then cured under heat and pressure (Fig. 3.4).

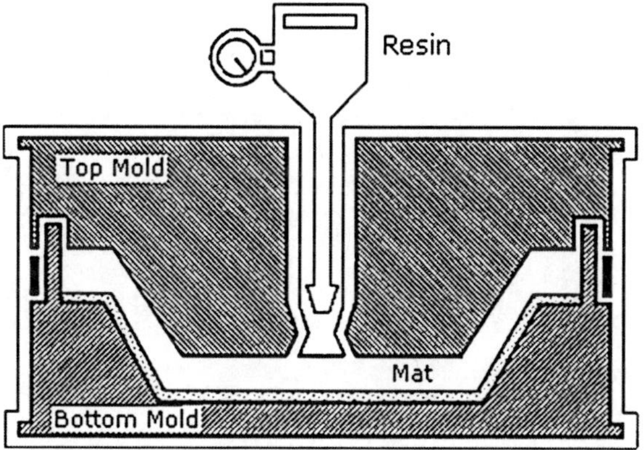


**Fig. 3.4** Pictorial view of RTM

Advantages:

1. very large and complex shapes can be made efficiently and inexpensively
2. production times are very short compared to layup
3. low clamping pressures
4. better surface definition than layup
5. inserts and special reinforcements are easily added
6. operators may be unskilled
7. A large number of mold materials may be used
8. part consistency is good
9. worker exposure to toxic chemicals is reduced.

Disadvantages:

1. mold design is complex
2. material properties are good, but not optimal
3. resin to fibre ratio is hard to control, and will vary in areas such as corners
4. reinforcement may move during injection, causing problems.

### *3.4.7 Pultrusion*

In the pultrusion process, continuous roving strands are pulled from a creel through a strand-tensioning device into a resin bath. The coated strands are then passed through a heated die where curing occurs. The continuous cured part, usually a rod or similar shape, is then cut to the desired length (Fig. 3.5).

Advantages:

1. good material usage compared to layup
2. high throughput
3. higher resin contents are possible.

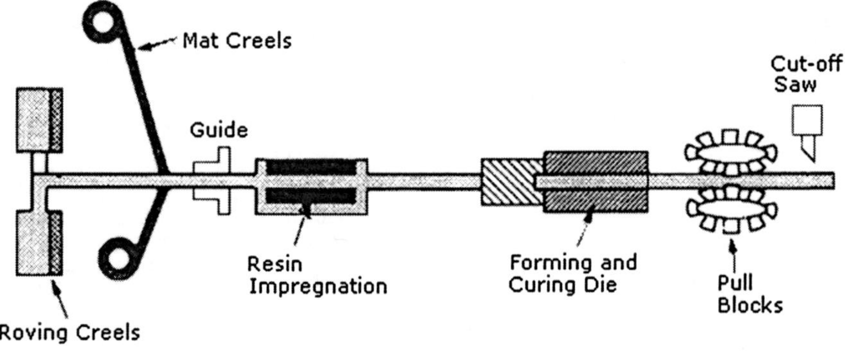


**Fig. 3.5** Schematic view of pultrusion

Disadvantages:

1. part cross section should be uniform
2. fibre and resin might accumulate at the die opening, leading to increased friction causing jamming, and breakage.
3. when excess resin is used, part strength will decrease
4. void can result if the die does not conform well to the fibres being pulled
5. quick curing systems decrease strength.

### *3.4.8 Vacuum Bagging, Autoclave Cure*

Most parts made by hand lay-up or automated tape lay-up must be cured by a combination of heat, pressure, vacuum, and inert atmosphere. To achieve proper cure, the part is placed into a plastic bag inside an autoclave. A vacuum is applied to the bag to remove air and volatile products. Heat and pressure are applied for curing. Usually an inert atmosphere is provided inside the autoclave through the introduction of nitrogen or carbon dioxide. Exotherms may occur if the curing step is not done properly.

Application of a vacuum to the resin helps eliminate residual materials/gas trapped in the uncured resin (Fig. 3.6).

## 3.5 Operating Variables Affecting WPC Microcellular Foams

Generally, wood-polymer composites are processed by either injection molding or extrusion. These equipment have many operating variables like screw speed, screw design, L/D ratio, temperature profile, etc., which affects the foam characteristics of WPC. Here we give a brief discussion on the effect of operating variables on

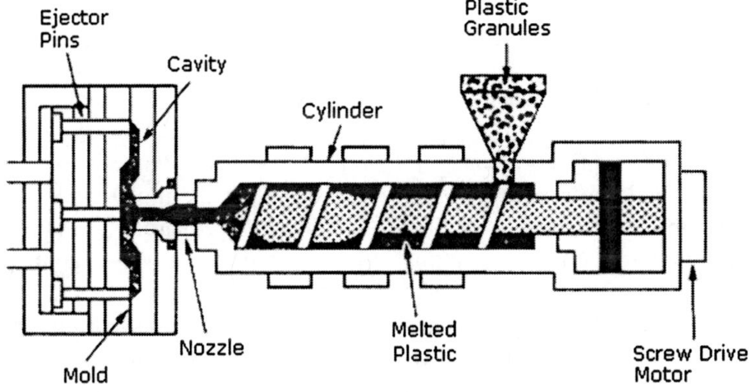


**Fig. 3.6** Pictorial view of vacuum bagging, autoclave cure

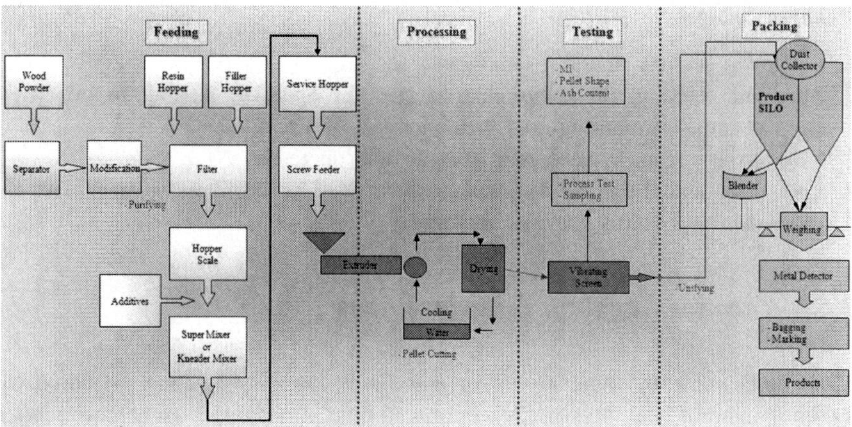


**Fig. 3.7** Schematic presentations of finished WPC composites from raw materials

foam characteristics. Our discussion will be restricted to extrusion and injection molding, since the variables affecting batch foaming has been extensively discussed in Chap. 7. A brief description about the preparation of the WPC composites from the raw materials to the finished product have been depicted below in a schematic view (Fig. 3.7).

### *3.5.1 Injection Molding*

Injection molding is one of the most widely used techniques in the wood-polymer composite industry, in which process parameters setting and optimization are recognized as one of the important approaches to improve the quality of the molded parts and shorten the cycle time. Generally, optimal process parameters are difficult to achieve because the process is affected by a large number of interrelated factors that relate to the molding machine, molding material, and mold structure. Injection molding belongs to the category of large-scale system, involving many fields of science and engineering, such as rheology, heat transfer, fluid dynamics, friction, polymer science, and control theory (Fig. 3.8).

There are too many interrelated and complex process parameters involved in the injection molding process, all of which can affect the molded parts quality. Moreover, the relations among these parameters are hardly possible to be expressed explicitly. Therefore, setting and optimization of process parameters for injection molding is a highly skilled task and is based on the operator's "know-how", experience and intuitive senses, which can only be acquired through long-term practice. The multitude numbers of variables that need to be optimized in a injection molding process are usually very complex and involve many personnel. A flow chart of this optimizing is shown in the following figure. This technique is

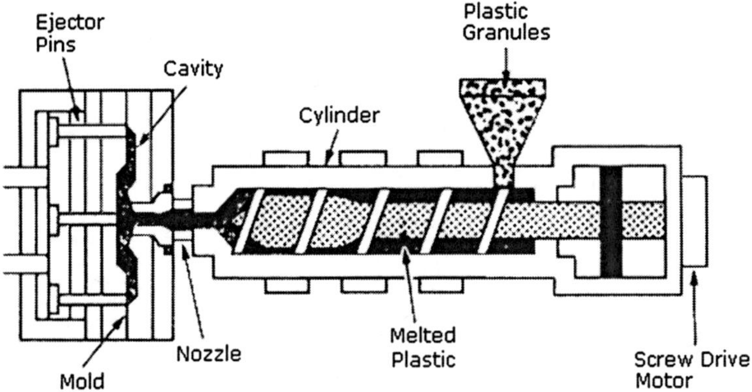


**Fig. 3.8** Schematic views for injection molding

called FALLO procedure, Follow all opportunities. This is a old technique and is being replaced by such techniques based on Fuzzy logic and Artificial neural networks. The advantages of newer techniques are there is a window of flexibility in the sense that "minimization" of the problems is allowed rather than complete "elimination" of the problems (which a seasoned injection molder knows its virtually impossible to run a trouble free injection molding or extrusion facility) (Fig. 3.9).

## 3.5.2 Extrusion

The extruder is the heart of the WPC extrusion system, and the primary purpose of the extruder is to melt the polymer and mix the polymer, wood and additives in a process referred to as compounding. In addition, the extruder conveys the compounded wood-polymer mixture through the die. There are four primary types of extrusion systems used to process WPC lumber. These are the (1) single screw, (2) co-rotating twin screw, (3) counter-rotating twin screw, and (4) Woodtruder™. Cost for an extruder can vary from $150,000 for a simple single screw extruder to over $1 million for a complete wood plastic composite lumber extrusion system [1].

### 3.5.2.1 Single Screw Extruder

The single screw fiber composite extruder is the simplest extrusion system for producing WPC lumber. A typical single screw extruder will have a barrel length to diameter (L/D) ratio of 34:1. It will employ two stages, melting and metering, and a vent section to remove volatiles. The material form for the single screw ext ruder will be pre-compounded fiber filled polymer pellets. A dryer may also be

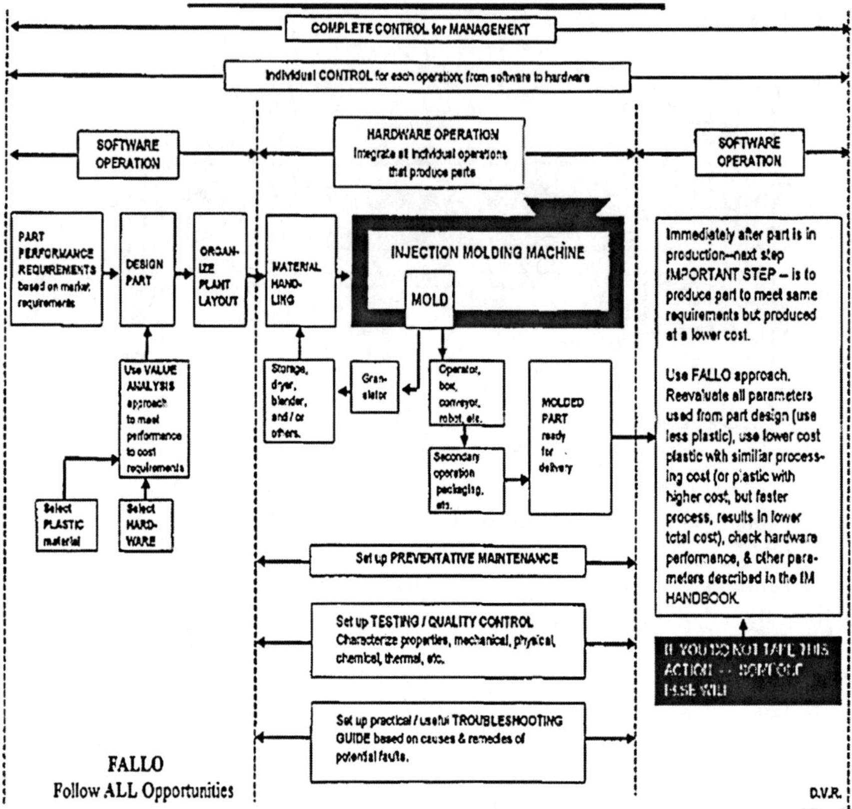


**Fig. 3.9** FALLO procedures to optimize injection molding technique

required to dry the pellets. The material feed method is usually by gravity hopper. The melting/mixing mechanism is barrel heat and screw shear.

Advantages of the single screw extruder are it's a proven technology and has the lowest capital acquisition cost. Disadvantages include: high raw material cost, lower output rates, drying system required; polymer is melted with the fiber with greater risk of fiber thermal decomposition, high screw rpm with greater risk of burning at the screw tip, and inability to keep melt temperature low with higher head pressures.

#### 3.5.2.2 Counter-Rotating Twin-Screw Extrusion

Counter rotating twin-screw extruders excel in applications where heat sensitive polymers like rigid PVC are utilized, low temperature extrusion for fibers and foams, non-compounded materials like powder blends, materials that are difficult to feed, and those materials that require degassing. The counter rotating twin screw can either have parallel or conical screw configurations. The fiber/flour and

**Fig. 3.10** Pictorial views for typical extruder for WPC work

polymer are in the same polymer size, usually 40 mesh. Material preparation includes fiber drying followed by high intensity blending with the polymer and additives. The material feed method usually utilizes a crammer feeder. The melting/mixing mechanism is barrel heat and screw mixing. Screw mixing is accomplished through screw flight cut-outs and gear mixers. Moisture removal is through vacuum venting.

Advantages of counter-rotating twin screw extrusion include its low screw rpm and low shear mixing and it is a proven technology. Disadvantages include that a drying system is required, a size reduction system for fed materials may be necessary, a pre-blending system is required, material transportation can impact the mix feed ratios. Because of the need for a dryer, additional plant floor space is required, higher operational costs including power maintenance, and labor. The polymer is melted with the fiber with a greater risk of burning the fiber (Fig. 3.10).

#### 3.5.2.3 Co-Rotating Twin Screw and Hot Melt Single Screw Wood Composite System

A co-rotating twin screw in combination with a hot melt single screw can be used to produce wood plastic composite lumber. In this case, a parallel 40:1 L/D co-rotating twin screw extruder is coupled with a "hot melt" 10:1 L/D, single screw extruder [2]. The material for this system is wood flour or fiber at ambient moisture content (5–8%) and the polymer and additives can be in their natural states. No material preparation is required in terms of pre-blending components. The preferred material feed systems are gravimetric feeders and twin-screw side feeders. The melting/mixing mechanism includes barrel heat, screw rpm and screw mixing. Moisture removal is accomplished through the use of atmospheric and vacuum vents. Advantages of this system include the ability to process wood at ambient moisture content since the extruder is used to dry the fiber with the elimination of drying and pre-blending operations, and good fiber/polymer mixing. Disadvantages include the need for peripheral feeding systems, high screw rpm and no screw cooling (greater risk of burning), inability to keep melt temperature low with higher head pressures, and polymer is still melted with the fiber (greater risk of burning, more difficult to vent).

#### 3.5.2.4 Woodtruder™

The Woodtruder™ includes a parallel 28:1 L/D counter-rotating twin- and a 75 mm single-screw extruder, a blending unit, a computerized blender-control system, a die tooling system, a spray cooling tank with driven rollers, a traveling cut-off saw, and a run-off table [1]. As processing begins, ambient moisture content wood flour is placed into the unit's fiber feeder and dried within the twin screw. Meanwhile, separate from the fiber, the plastics are melted. The melting/mixing mechanism includes barrel heat and screw mixing. The separation of wood conveying and plastic melting ensures that fibers will not be burned during plastic melting and that the melted plastic will encapsulate the fibers completely. These materials are then mixed, and any remaining moisture or volatiles are removed by vacuum venting.

Advantages of this system include that the flour and additives are in their natural states and no material preparation is required. Gravimetric feeders are preferred as material feed method. Advantages of the Woodtruder™ include the ability to process fiber at ambient moisture content (5–8%), separate melting process of the polymer, good polymer/fiber mixing, screw cooling is included on the twin screw, the ability to maintain a low melt temperature with a high head pressure, superior venting, the elimination of drying, size reduction and separate pre-blending equipment, highly flexible integrated process control system for material feeding and extruder unit operations (Fig. 3.11).

#### 3.5.2.5 Miscellaneous Post-Extruder Unit Operations

Along with the extruder, the die is an important part of the WPC lumber extrusion system. The die dictates the dimensions and profile (shape) of the extruded part. The die is typically heated using band or cartridge heating elements, and may employ air-cooling to adequately process hollow profile parts. Dies can be quite

**Fig. 3.11** Pictorial views for Woodtruder™ for WPC work

simple or complex depending on the desired profile. The costs for dies can range from $15,000 for a simple die up to $50,000 or more for a foaming or co-extrusion die [3, 4]. After the die, comes the cooling tank, which is used to "freeze" the extruded profile in its linear shape. The cooling tank consists of a conveyer system with water spray heads that spray cool water on the profile extrudate. The cooling tank may be 20–40 feet long depending on the extruder material output and the cooling capacity required. The water spray is typically recycled and may go through a chiller or heat exchanger to keep the spray water cool. After the cooling tank the WPC profile goes through a cut-off saw that can cut the lumber to the desired lengths.

### *3.5.3 Rheotens*

Though the investigation of elongational behavior of polymer melts originated in 1970s, there are still many gaps between the research and industrial requirements. A common experiment is the so called Rheotens experiment. The rheotens experiment developed by Wagner is a simple and practical method almost similar to commercial melt flow index experiments. A schematic representation of the rheotens test is shown in the following figure (Fig. 3.12).

In Rheotens experiment the extrudate strand produced in either extruder or injection molding machine is drawn with the help of rollers at a predetermined speed (called as take up speed). This take up speed is gradually increased until the polymer strand breaks. A plot of draw down force $F$ is plotted as a function of drawdown speed. A typical Rheotens plot is shown in figure below (Fig. 3.13).

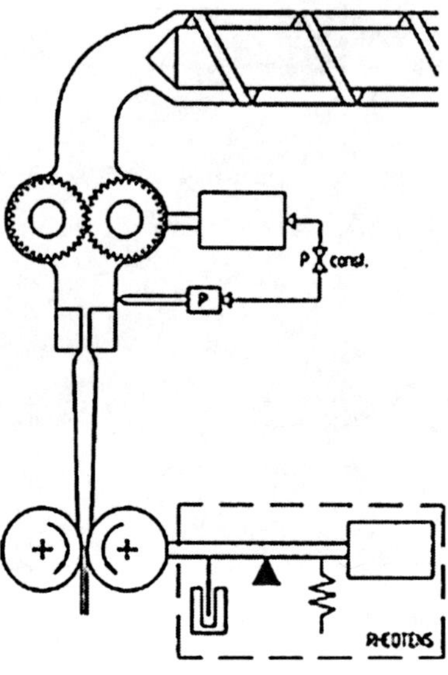


Fig. 3.12 Schematic diagram of rheotens equipment fitted to an extruder

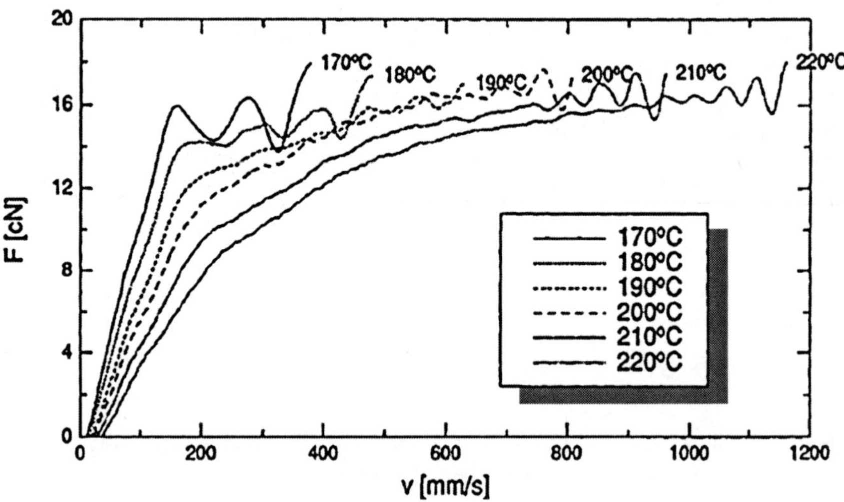


Fig. 3.13 Rheotens plot of HDPE at different temperatures

Rheotens plot are relevant in foaming than steady state elongational experiments (as in Malvern or Monsanto rheometers) used typically in rubber or thermodynamic processing. Typically during bubble growth the cell walls experience biaxial stretching, the same mode of deformation as in other thermoplastic processing like (thermoforming, film blowing, blow molding, etc.). But, in addition to this biaxial stretching, two other types of starching like uni-axial and planar also occurred.

## References

1. Gardner, D.J., Murdock D.: Extrusion of Wood Plastic Composites. http://www.entwoodllc.com/PDF/Extrusion%20Paper%2010-11-02.pdf (2010)
2. ASTM: Standards boost an industry: recycled plastic lumber gains ground, pp. 22–26. ASTM Standardization News, July 1999
3. Lu, J.Z., Wu, Q., McNabb Jr, H.S.: Chemical coupling in wood fiber and polymer composites: a review of coupling agents and treatments. Wood Fiber Sci **32**(1), 88–104 (2000)
4. Rowell, R.M., Lange, S.E., Jacobson, R.E.: Effects of moisture on aspen-fiber/polypropylene composites. In: Progress in Woodfibre-Plastic Composites Conference 2002, Toronto, Canada

# Chapter 4
# Recent Past about WPC Work

## 4.1 Wood–fiber/Plastic Composites (WPCs)

In recent years, significant efforts have been made to manufacture thermoplastic composites using such natural fibers as wood sawdust, wheat straw, nut shell fiber, and jute fiber [1–3]. The rationable behind these efforts is that the use of natural fibers offers several benefits, including low cost, high specific properties, renewable nature, and biodegradability. Wood fibers are the most favored form of fibers in commercial usage. Because of their high specific stiffness and strength, Wood–fiber/Plastic Composites (WPCs) are a cost-effective alternative to many plastic composites or metals [4]. Wood fiber is a non-abrasive substance, which means that relatively large concentrations of this material can be incorporated into plastics without causing serious machine wear during blending and processing. In spite of their higher price, WPCs are becoming increasingly acceptable to consumers as a replacement for natural wood due to such advantages as durability, color permanence, resistance to degradation and fungal attacks, and reduced maintenance. Furthermore, adding wood fibers to plastic products makes good use of waste wood. WPCs are mainly employed in building products, such as decking, fencing, rails, door and window profiles, and decorative trims. Moreover, these composites are also gaining acceptance in automotive and other industrial applications [5].

Polyolefins such as PE and PP are the most commonly used polymer matrix for WPCs because of its relatively low processing temperature and good processability [6]. However, the low compatibility between the hydrophilic wood fiber and hydrophobic polymer matrix is one of the major reasons for limited use of wood fiber as reinforcement. In many cases, poor mechanical and physical properties of a wood fiber composite can be attributed to the weak fiber-matrix interfacial bond caused by the low compatibility. Obviously, how to improve the compatibility between the two components is a key to success in the area. The surface modification of fibers or use of external processing aids can facilitate the dispersion and adhesion of these fibers in the polymer matrix [7]. The most widely used coupling

J. K. Kim and K. Pal, *Recent Advances in the Processing of Wood–Plastic Composites*, Engineering Materials, 32, DOI: 10.1007/978-3-642-14877-4_4,

agents (based on reactive groups) are derivatives of either maleic anhydride or siloxanes. It has been shown that the mechanical properties of WPCs are improved significantly when coupling agents are used in the composites [8–11].

Another disadvantages of using wood fiber as filler is their high water absorption or desorption when subjected to changes in the relative humidity of the environment. So the moisture that is present in wood fibers has many adverse effects during processing and also on final products. It causes undesirable voids in solid WPCs, and dimensional variations. It also results in poor adhesion due to its sorption to the fibers, and consequently lowers the mechanical properties [12].

## 4.2 Flammability of WPC

The third critical drawback of WPCs is their high flammability. Improving their flame retardancy will thus expand the range of their applications. Halogenated flame retardants, such as organic brominated compounds, are often used to improve the flame-retarding properties of polymers; unfortunately, these also increase both the smoke and carbon monoxide yield rates due to their inefficient combustion [13]. The other commonly used flame retardants are aluminum trihydrate (ATH), magnesium hydroxide [$Mg(OH)_2$], ammonium polyphosphate (APP) or intumescent systems. However, they all exhibit some significant disadvantages. For example, the application of ATH and APP requires a very high loading of the filler (40–60 wt%) within the polymer matrix to obtain acceptable performance levels, which yields high-density products and also adversely influences mechanical properties and processability [14, 15].

As a component of flame-retardant systems material, silica has been applied to several polymers such as polypropylene, poly(methyl methacrylate) (PMMA), epoxy resin, poly(ethylene-co-vinyl acetate) (EVA), butadiene-acrylonitrile rubbers. The effect of silica gel structure on the flammability properties of polypropylene has been investigated by Gilman et al. [15]. They studied the effect of three silica gels with different pore volume, particle size and surface silanol concentration. Cone calorimeter tests revealed the dramatic effect of silica gel pore volume on the heat release rate (HRR) of PP containing 10 wt% in silica. The incorporation of high pore volume silica led to a marked reduction in the HRR. There was no noticeable effect of particle size on the flammability properties but a significant effect of the surface silanol concentration was observed. In another study, Kashiwagi et al. [16] also investigated the performances of various types of silica, silica gel, fumed silica and fused silica as flame retardants in non-char-forming thermoplastics (e.g., polypropylene) and polar char-forming thermoplastics (e.g., polyethylene oxide). The addition of low density, large surface area silica, such as fumed silica (140 and 255 $m^2/g$) and silica gel (400 $m^2/g$) to polypropylene and polyethylene oxide significantly reduced the HRR and mass loss rate.

Sain et al. [17] found that magnesium hydroxide can effectively reduce the flammability (almost 50%) of natural fiber filled polypropylene composites. No synergetic effect was observed when magnesium hydroxide was used in combination with boric acid and zinc borate, marginal reduction in the mechanical properties of the composites was found with addition of flame retardants. Zhao et al. [18] reported the mechanical properties, fire retardancy and smoke suppression of the silane-modified WF/PVC composites filled by modified montmorillonite (OMMT), and observed that the fire flame retardancy and smoke suppression of composites were strongly improved with the addition of OMMT. Li et al. [19] investigated the mechanical property, flame retardancy and thermal degradation of LLDPE-wood fiber composites, and found APP decreased initial temperature of the thermal degradation, and promoted char formation of the composite.

## 4.3 Polymeric Foams

### *4.3.1 Introduction of Polymeric Foams*

Polymeric foams [20–22] defined as materials consisting of gases voids surrounded by denser polymer matrix, have attracted enormous research attention due to their wide applications in insulation, cushion, absorbent, and weight-bearing structures. More recently, foams with interconnected pore structures have been applied as the tissue engineering scaffolds for cell attachment and growth. A variety of polymers has been used to synthesize polymeric foams. Based on different standards, there are several ways to classify polymeric foams. According to the original polymers, polymeric foams can be classified into two categories: reprocessable thermoplastic foams and un-reprocessable thermoset foams.

Another standard is based on the size of foam cells. Polymeric foams can be classified as macro cellular foam (>100 μm), microcellular foam (1–100 μm), ultra microcellular foam (0.1–1 μm), or nano cellular foam (<0.1 μm).

With respect to different material composition, foam morphology, physical properties and thermal characteristics, polymeric foams can also be classified into rigid or flexible foams. Rigid foams have a wide range of applications such as building and construction, transportation, floatation and cushion, packaging, molding and food/drink containers. On the other hand, flexible foams can be used for bedding, textile, shock and sound attenuation, gaskets, and sports applications. Regarding different cell morphologies, polymeric foams can also be defined as either closed cell or open cell foams. By selecting foaming materials and foaming processes, the formation of open cells or closed cells can be controlled. In closed cell foams, each cellular structure is surrounded by a complete cell wall and all cells are separated. In open cell foams, all cells are virtually connected to each other in the absence of the cell walls. Moreover, the foam structure is supported by the struts and ribs instead of the cell walls. Compared to closed cell foams, open cell foams usually exhibit higher absorptive capacity, higher permeability and better sound damping ability. However, the open-channel structure will lead to less efficient insulation behaviors.

Various techniques can be used to produce polymeric foams. For large-scaled production, the utilization of blowing agents is the most practical method. For thin film foams, other methods such as phase inversion, leaching [23], and thermal decomposition are commonly used [24].

A typical formulation of a foaming system is composed of the polymer (or polymer monomer), blowing agent, nucleating agent, and other necessary additives (fire retardant, surfactant, catalyst, etc.)

## *4.3.2 Blowing Agents*

Generally, there are two types of blowing agents: physical blowing agents (PBA) and chemical blowing agents (CBA). CBAs are usually reactive species that can produce gases by certain chemical reactions or thermal decomposition. CBA are substances that decompose at processing temperatures, thus liberate gases like $CO_2$ and/or $N_2$. Solid organic and inorganic substances (such as azodicarbonamide and sodium bicarbonate) are used as CBA. In general, CBA are divided by their enthalpy of reaction into two groups including exothermic and endothermic foaming agents. The reaction that produces the gas can either absorb energy (endothermic) or release energy (exothermic). Nowadays, a combination of exothermic and endothermic CBA is also used for foaming. Requirements to an ideal chemical foaming agent (CBA) [25] are:

Decomposition reaction has to be in a defined temperature range, according to used polymer.

- Avoid very fast reaction (explosion) by decomposition of the CBA.
- Prevention of a heat build-up or burning.
- Easy mixing and uniform dispersion of the CBA in the polymer.
- High gas yield and feasibility of the CBA.
- CBA should not be corrosive for tools.
- No discoloration and plate out of the polymer.

PBA are materials that are injected into the system in either a liquid or gas phase. Some PBA such as, pentane or isopropyl alcohol, have a low boiling point and remain in a liquid state in the polymer melt under pressure [22]. When the pressure is reduced, the phase change of the foaming agent from liquid to vapor happens immediately and the vapor comes out of the solution with the polymer, thereby expanding the melt. As the boiling point of a gas lowers, the volatility of the gas increases. Higher volatility or vapor pressure requires more pressure to keep the gas in its liquid phase in the polymer melt. Another type of PBA is inert gases, such as $N_2$ or $CO_2$. Inertness represents the reactivity and corrosiveness of a gas to the polymer, any additive, the machinery, or the surrounding environment. As a blowing agent becomes more inert, it is less reactive (or corrosive) to its surroundings. These inert gases dissolve as vapors in the polymer melt and diffuse out of the solution as vapors to expand the polymer melt. The solubility of the gases affects the final density. Both $CO_2$ and $N_2$ have low solubility in polyolefin matrices. The diffusivity of these gases is important in maintaining the cell structure and resulting density.

In conventional foam processing, the most commonly utilized blowing agents are FCs, CFCs, *n*-pentane, and *n*-butane [26]. These blowing agents have a high solubility, and can thus be dissolved in large quantities into the polymer matrix. For example, the solubility of FC-114 in polystyrene is above 20 wt% at a pressure of 6.9 MPa and at a temperature of 200°C [27]. These blowing agents allow for a foam structure that have high void fraction at low processing pressure and high volume expansion due to small amounts of gas loss. Since these blowing agents have low diffusivities due to their larger molecular size, the loss of gas from the extrudate during expansion is small [27]. Therefore, the final foam product can have a low foam density. Despite all the advantages of the conventional foaming blowing agents, some serious environmental and safety concerns exist. The use of CFCs was stopped by the Montreal Protocol [28] signed by 24 countries in 1987. Other blowing agents, such as *n*-pentane, are hazardous because of their high flammability. As a result, alternative blowing agents for polymer foam processing are being researched and developed. Of these, $CO_2$ is the most favorable choice due to its environmentally benign properties.

### *4.3.3 Procedure*

As shown in Fig. 4.1 [29], the basic foaming process consists of three main steps: formation of polymer/blowing agent solution/mixture, cell nucleation, and cell growth. To ensure that all cells are created by nucleation, good mixing is necessary to form a homogeneous solution composed of the foaming agent and polymer melt. Next, cell nucleation, a phase separation phenomenon, is induced by a thermodynamic instability to form nuclei with the critical size. Such instability is usually a temperature increase or pressure decrease, which can reduce the solubility.

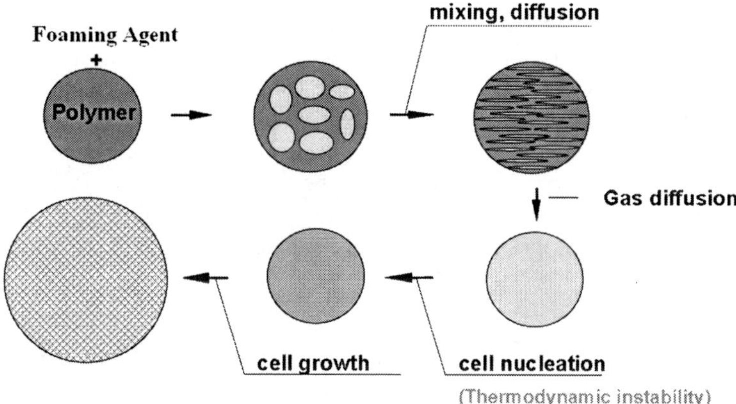


**Fig. 4.1** Diagram of the foaming process

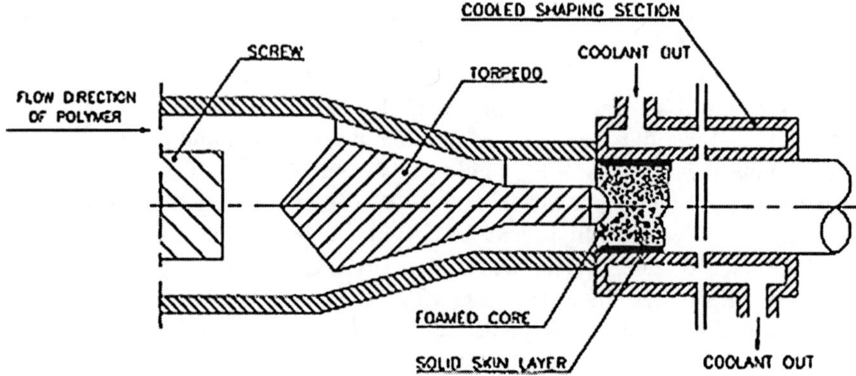


**Fig. 4.2** Schematic representation of Celuka process

Compared with the cell nucleation, determining the cell density, cell growth, starting with the critical size, decides the cell size. Cell growth is a complex procedure controlled by fluid dynamics, mass and heat transfer, and polymer rheology.

#### 4.3.3.1 Conventional Continuous Foaming

A good review of conventional plastic foams and their processes is provided in some literature [20, 26]. There are basically two main categories of producing plastic foams by conventional foaming process.

- Controlled foam extrusion process.
- Free foaming extrusion.

In controlled foam extrusion process, the schematic diagram of which is shown in Fig. 4.2 [30], the polymer is melted in an extruder followed by injection of blowing agent which mixes with the polymer matrix and dissolves in it. This polymer melt containing blowing agent is subsequently forwarded to the head of an extruder which is partially plunged with a torpedo or mandrel. This mixture is rapidly cooled in a water cooled shaper connected to the head of an extruder. The foaming process in this procedure proceeds from inwards towards the outer layer. Therefore the outer skin layer has less or no microcells (which is cooled more rapidly than the inner core). In this process both cell nucleation and cell growth occurs within the shaper. The above procedure is commercially available as "Celuka Process" [31] and is widely used in industry. However there are many disadvantages of this Celuka process.

- First is the formation of solid or less foam outer skin layer which needs to be cut thereby leading to wastage of raw materials. However this problem can be overcome by an optimal design of the die.
- Secondly since the foaming process starts from core to outwards, the cell density and cell distribution at core seems to be higher when compared to skin.

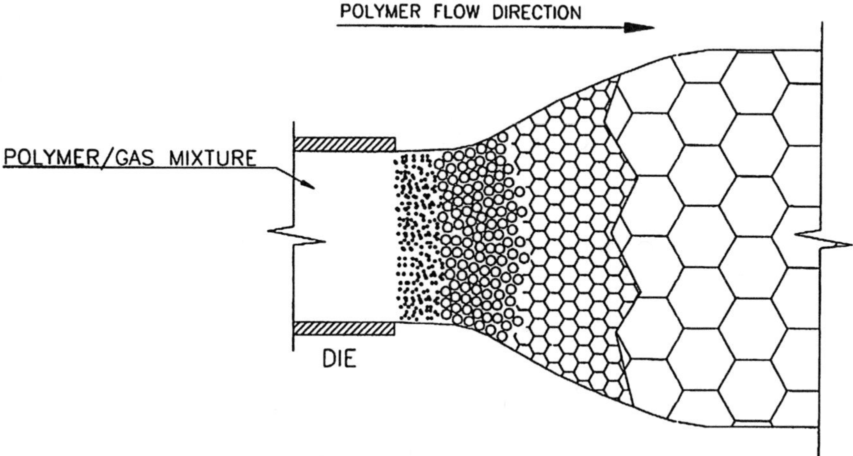


**Fig. 4.3** Free foaming extrusion processes

This problem can be overcome by use of special type of screws. (A detailed discussion of screw characteristics is given in the later parts of this chapter).

In contrast to controlled foam extrusion, free foaming extrusion as the name itself indicates the foaming of wood/polymer mixture is allowed to expand freely without any constraints. A schematic representation of this system is shown in Fig. 4.3. The rate controlling step in this process is the temperature of the melt at the die exit. If the temperature of the melt is too high, there will be a sudden burst of bubbles within the polymer matrix, thereby leading to open cell structure. On the other hand, if the temperature is low enough there will not be sufficient foaming. So, some researchers tried gradual reduction of temperature but it substantially slows down the whole process thereby affecting productivity.

In a conventional foam process, foam products of various densities can be produced. However, the state-of-the-art foams have a fully grown cell size greater than 10 μm, a cell population density lower than $10^6$ cells/cm [3, 20, 26] and a non-uniform cell size distribution. Hence, the mechanical properties of the conventional foams are poor.

#### 4.3.3.2 Batch Foaming

In the batch foaming process (Fig. 4.4) [32–35], pre-shaped samples are placed in a pressurized autoclave and saturated with the foaming agent ($CO_2$, $N_2$, etc.) at a selected saturation temperature and pressure. The saturation time varies from several hours to several days according to the gas diffusivity and the sample dimension. After saturation, there are two choices possessing different driving forces for cell nucleation.

If the saturation temperature is relatively high [near or above the glass transition temperature ($T_g$) of the sample], this saturation temperature can be treated as the

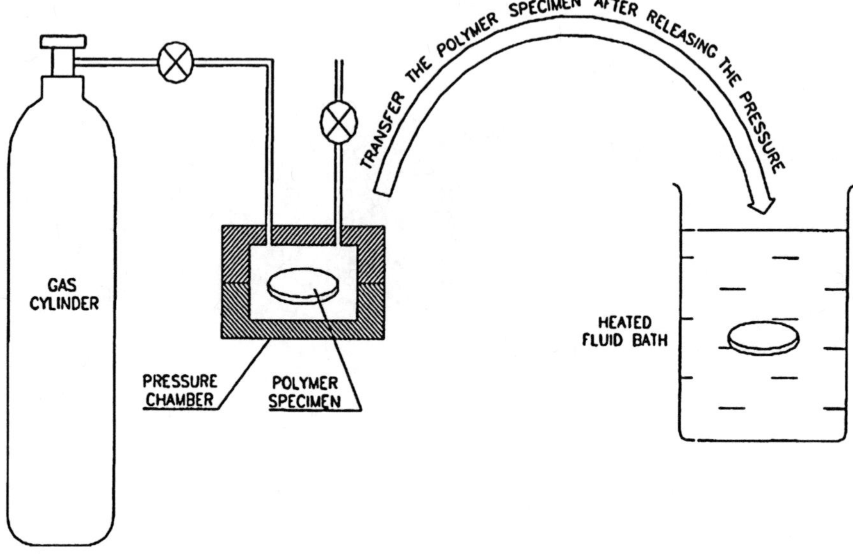


**Fig. 4.4** Microcellular foaming by a batch process

foaming temperature. The sample can be foamed at such a temperature when the pressure is rapidly released. Here, the nucleation is induced by the pressure drop ($\Delta$P) and pressure drop rate ($\Delta$P/$\Delta$t). Otherwise, if the saturation temperature is very low and far below $T_g$, the sample cannot nucleate when the pressure is released due to the rigidity of the polymer matrix. The saturated sample can be then dropped in a high temperature water (oil) bath to get foamed. This high temperature is the foaming temperature; and correspondingly, the nucleation driving force is the temperature increase ($\Delta$T) and temperature increase rate ($\Delta$T/$\Delta$t). The advantage of the second choice is that the gas solubility is high at the low saturation temperatures. Combinations of the two choices can be also used to nucleate samples.

#### 4.3.3.3 Semi-Continuous Foaming

Kumar and Schirmcr [36, 37] developed a semi-continuous process for production of solid state microcellular PET foams. In this process, layers of plastic sheet are interleaved with gas permeable materials (such as gauze porous paper sheet, etc.). The roll of the interleaved plastic and gas permeable material is saturated with an inert gas in a high pressure vessel. The plastic sheet is taken out from the vessel, unrolled, and separated from the gas permeable material. The saturated sheet is drawn through a heating station and resides sufficiently to achieve a desirable foam density. In this process, the gas tends to escape from the sheet after its removal from the vessel. Therefore, the foaming must be processed promptly to avoid excess gas loss. In other word, only a finite length of the sheet can be foamed at one time, and hence, the process is called "semi-continuous".

#### 4.3.3.4 Continuous Extrusion Foaming

Many studies have been carried out in polymer extrusion process by using CBA [38–44] or PBA [45–48]. A schematic of the extrusion foaming process is shown in Fig. 4.5. Compared with the batch process, a continuous extrusion foaming process is more economically favorable because of its high productivity, easy control, and flexible product shaping. By producing foam continuously, the foam extrusion process is strict because of the mass, momentum, and energy transfer required for a steady flow field. The pellets of polymer are mixed with a blowing agent and are melted by heating at high pressure in an extruder. The blowing agent generates the gas bubbles, resulting from chemical reactions at the decomposition temperature of the blowing agent. At that time, the generated gas bubbles are dissolved in the polymer melt under the influence of the pressure of the inner part of the extruder. When this mixture is exposed to the atmosphere through the die of the extruder, it reaches the super saturation state, the pressure of the polymer melt decreases, and the dissolved gas is able to form the nucleus. This nucleus grows by the diffusion of dissolved gas from the polymer melt and by the momentum transfer, and expands to the steady state of the concentration of gas dissolved in the polymer melt. Because of the cooling at the surface of the mold, the temperature of the polymer melt decreases, and the decrease in temperature of the polymer melt causes the viscosity of the polymer melt to increase. The gas bubbles near the surface are hindered from growing. Finally, the polymer melt near the surface is solidified and the polymer melt in the central part is foamed. The final foam is the type of structural foam with few gas bubbles at the surface. The decrease of the density of polymer resin is the function of the concentration of the dissolved gas, the kind of polymer, and the process variables.

A traditional extrusion process is suitable for foam manufacture after a few modifications, such as a venting port for $CO_2$ injection, a special mixing element for creating a homogeneous solution, and a specific foaming die. The extrusion foaming can be performed on a single screw extruder, a twin screw extruder, or

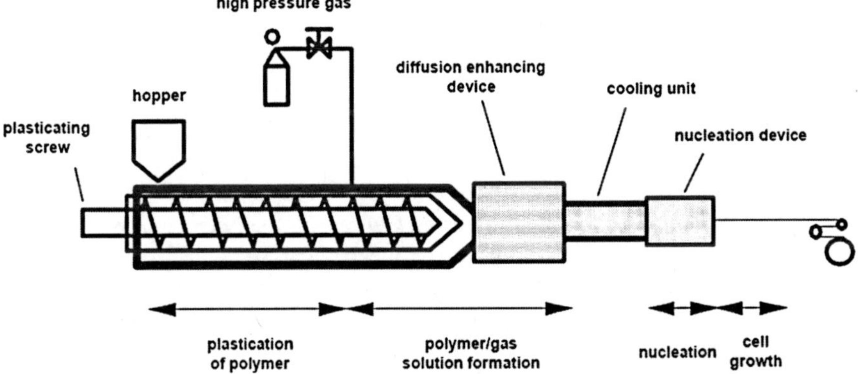


**Fig. 4.5** Schematic of a single foaming extruder

tandem extruders where two extruders are connected together and the mixing and temperature cooling control are independent procedures on each extruder.

A typical foaming extrusion [49, 50] begins with the plasticization of polymer resin after which $CO_2$ is injected into the extruder barrel. A decreasing temperature profile from the hopper to the die is usually applied for the foaming extrusion. High temperatures from the hopper to the gas injection port ensure the complete melt of the polymer resin. Once the gas in injected in the barrel, low barrel temperatures can be applied due to the viscosity reduction. A low foaming temperature (or die temperature) is always favorable to get good foam structure. Mixing elements, such as static mixers, can be attached to the extruder to further improve the mixing effect and pre-control the melt temperature. A homogeneous single-phase solution consisting of the polymer melt and the foaming agent is created by the screw rotation and the on-line mixers.

The rapid, large pressure drop through the die induces cell nucleation, although a device to create a sudden temperature increase in the extrusion die was also reported [51]. Again, the purpose of the pressure drop or temperature increase is to reduce the gas solubility in the polymer melt and therefore induce the phase separation. After the nucleation die, a shaping die is usually necessary to control the product shape and the foam expansion. Once the extrudate temperature is lowered below its $T_g$, the foam structure is vitrified and the foam sample can be collected.

#### 4.3.3.5 Injection Molding Foaming

Foam injection molding is another commonly used polymer processing technique that combines gas dissolution, cell nucleation, and cell growth, with product shaping. Compared with the extrusion foaming process, foam injection molding has its own advantages. For example, it is convenient to produce parts with complex geometry. By choosing the right size of the injection nozzle or mold gate, the pressure drop (or pressure drop rate) through them can be very high, which provides the high thermodynamic instability for cell nucleation.

Injection foam molding using CBAs has already applied to much kind of polymers. Guo et al. [52] studied the cell structure and dynamic properties of injection molded polypropylene foams, the results showed that with the addition of nanoclay as the nucleating agent, more uniform cell structure with high cell densities and low cell dimensions were obtained and the cell structure could be improved greatly by optimizing key parameters such as shot size, CBA concentration, back pressure, injection speed, and melt temperature.

Currently, foam injection molding using $CO_2$ as the foaming agent is applied to produce lightweight products with strong mechanical strength. When one considers the preservation of polymer molecular structure at high melt temperature, $CO_2$ is a better choice than the thermal degradable CBAs. Applying MuCell molding technology invented by Suh et al. [53], Trexel [54] has successfully commercialized this technique to injection mold microcellular foams. Xu et al. [55] used injection molding to make microcellular foam after a series of modifications on certain components of a standard reciprocating-screw injection molding machine, such as the plasticizing unit, injection unit, hydraulic unit, clamping unit, and gas delivery unit. To successfully produce microcellular foams a new screw designed for better mixing and a new sealed barrel with gas injectors were used. The injection unit requires a fast injection speed to get the high pressure drop rate. It was found that a finer cell structure and more uniform cell size distribution can be achieved by controlling the pressure drop rate at the mold gate than at the injection nozzle. Of course, the injection speed should be controlled below the shear limit to prevent the melt fracture. Generally, foaming injection molding achieves increased melt flowability, lower injection pressures, faster cycle times, and greater dimensional stability and weight savings in molded parts [56, 57]. The reduced viscosity after the addition of supercritical $CO_2$ allows faster injection speed, lower injection pressure, and lower clamp tonnage. The shot size for the microcellular foam process is usually smaller than that for the solid molding process, which brings shorter recovery time. The pack and hold time is eliminated due to the internal gas pressure and the significantly less mass of material that needs to be cooled. The uniform cell distribution and expansion allow improved dimensional stability and diminish the surface flaws (ex. sink marks). However, to obtain a perfect surface finish, other techniques, such as a venting mold or co-injection, are needed. A new potential application of foaming injection molding is to make fiber filler articles [58]. The viscosity reduction after introducing the foaming agent can prevent the fibers to break during flowing in the mold and reduce the fiber orientation. Wong et al. [59] investigated the effects of processing parameters on the mechanical properties of injection molded thermoplastic polyolefin (TPO) foams. The findings showed that the mechanical

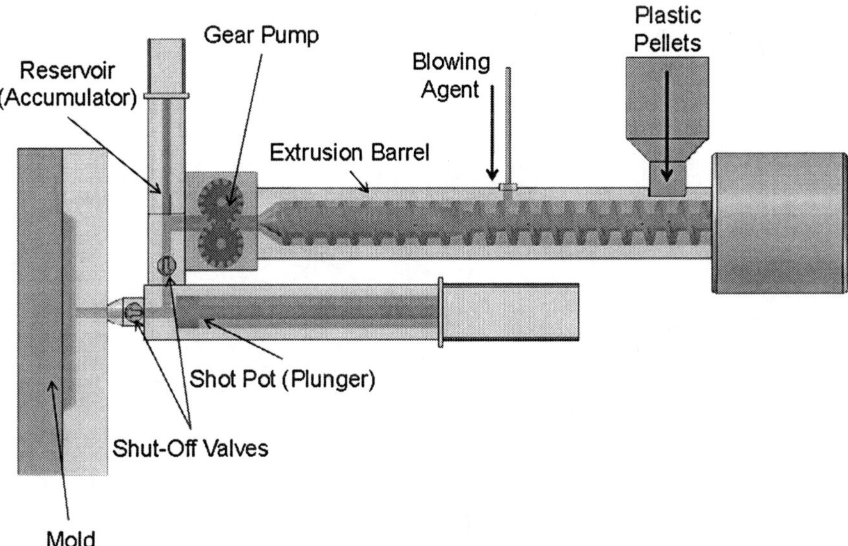


**Fig. 4.6** Schematic of advanced structural foam molding machine

properties were significantly affected by foam morphologies. The schematic of advanced structural foam molding machine was shown in Fig. 4.6.

## 4.3.4 Fundamentals in Polymeric Foaming

### 4.3.4.1 Formation of Polymer/Gas Solution

In continuous foam processing, it is essential to obtain the formation of a uniform solution since its quality significantly affects the number of bubbles nucleated later on in the process. Above all, the amount of blowing agent injected should be below the solubility limit of the processing pressure and temperature in order to ensure complete mixing and dissolving of gas into the polymer. This must be carefully controlled because large voids will form if an excess amount of blowing agent exists and cannot be dissolved into the polymer. For this reason, it is crucial to determine the solubility (or the amount) of blowing agent that can be absorbed and dissolved into the polymer at different processing temperatures and pressures. This information is necessary for the production of microcellular foam in order to avoid the presence of large voids.

### 4.3.4.2 Solubility

The solubility limit can be determined in a batch process over a limited range of temperature. In this process, a sheet polymer sample is saturated with the blowing agent by placing it in a high pressure chamber connected to the blowing agent reservoir. The blowing agent starts to diffuse into the polymer matrix, and diffusion continues until a certain limit of blowing agent concentration is reached (Fig. 4.7); however, theoretically, the limit occurs at time infinity. The instantaneous

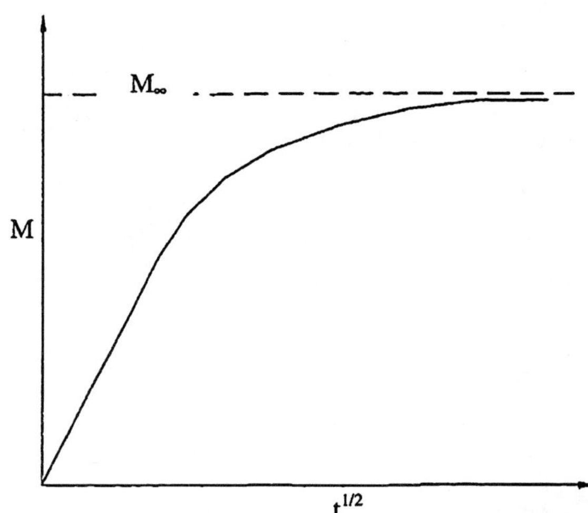


**Fig. 4.7** A typical uptake curve

concentration of the blowing agent in the polymer can be obtained using the following equation [60]:

$$\frac{M_t}{M_\infty} = 1 - \frac{8}{\pi^2}\sum_{m=0}^{\infty}\frac{1}{(2m+1)^2}\exp\left[\frac{D(2m+1)^2\pi^2 t}{h^2}\right] \tag{4.1}$$

where $D$ diffusivity (cm$^2$/s); $M_t$ mass uptake at time $t$, $g$; $h$ sheet thickness (cm); $M_\infty$ equilibrium mass uptake after an infinite time, $g$; $t$ elapsed time (s).

During the absorption process, the amount of mass uptake eventually tends to level off $M_\infty$. This maximum amount is related to the solubility limit of the gas in the polymer. The solubility limit can easily be calculated by dividing the mass uptake ($M_\infty$) by the mass of the polymer sample.

The maximum amount of gas that can be dissolved into the polymer (i.e., the solubility) depends on the system pressure and temperature and can be estimated by Henry's law [61]:

$$C_s = HP_s \tag{4.2}$$

where $C_s$ is the solubility of gas in the polymer in cm$^3$/g or $g_{gas}/g_{polymer}$, $H$ is Henry's law constant (cm$^3$ [STP]/g-Pa), $P_s$ is the saturation pressure in Pa.

The constant $H$ is a function of temperature described by:

$$H = H_0 \exp(-\frac{\Delta H_s}{RT}) \tag{4.3}$$

where $R$ is a gas constant in J/K, $T$ is the temperature in K, $H_0$ is a solubility coefficient constant (cm$^3$ [STP]/g-Pa), $\Delta H_s$ molar heat of sorption (J).

Here $\Delta H_s$, can be either a negative or positive value, depending on the polymer-gas system. Using Eqs. (4.2) and (4.3), one can determine the solubility of a blowing agent (such as $CO_2$) in the polymer at the processing pressure and temperature.

Wissinger et al. [62, 63] studied the sorption behavior of $CO_2$ in various polymers at relatively low temperatures (35, 50, and 65°C) and pressures up to 100 atm. Sato et al. [64, 65] characterized both solubility and diffusion coefficients of $CO_2$ and $N_2$ in polypropylene, high-density polyethylene, and polystyrene under high temperatures and a wider pressure range by using one gravimetric method. Zhang et al. [66] monitored the phase separation of the $CO_2$/PS system at 215°C by attaching an in line optical window on an extruder. They all presented a linear relation between the solubility and the saturation pressure, or a Henry's law behavior.

#### 4.3.4.3 Diffusivity

The initial slope of the curve in Fig. 4.7 corresponds to the diffusivity of the blowing agent into the polymer matrix. This slope can be used to calculate the diffusivity using the following equation [95]:

$$D \equiv \frac{0.04919}{\left(\frac{t}{h^2}\right)_{1/2}} \tag{4.4}$$

where $(t/h^2)_{1/2}$ in is the value of $(t/h^2)$ corresponding to $M_t/M_\infty = 1/2$. The time required for the completion of absorption can be approximated from Eq. (2.1) as given in the following:

$$t_D \equiv \frac{\pi h^2}{16D} \tag{4.5}$$

The diffusivity $D$ is mainly a function of temperature, and this can be represented as the following equation [60, 61]:

$$D = D_0 \exp\left(-\frac{E_d}{RT}\right) \tag{4.6}$$

where $D_0$ is the diffusivity coefficient constant in $cm^2/s$, and $E_d$ is the activation energy for diffusion in J. Thus, the diffusion rate can be increased by processing the plastic/gas mixture at a higher temperature. For instance, in the polystyrene (PS)-$CO_2$ system, $D_0$ and $E_d/R$ are 0.128 $cm^2/s$ and $4.35 \times 10^3$ K, respectively [96].

#### 4.3.4.4 Dissolution

In a continuous process, retaining an undissolved gas in the polymer matrix is possible if an excess amount of gas is injected. Therefore, it is critical to ensure that the amount of gas injected is below the solubility limit in the processing conditions. However, one advantage of using the extrusion foaming process is the reduction in dissolution time because of higher gas diffusivity at the high processing temperature. This makes the extrusion process a more cost-effective method. In addition to the proper amount of gas injection, a sufficient amount of dissolution time is required to generate a uniform solution. Although the appropriate amount of gas can be injected, it does not necessarily guarantee the formation of a uniform solution. If the required time of gas diffusion in the polymer matrix is longer than the melt residential time inside the system between gas injection and nucleation, it is obvious that a uniform solution would not be achieved.

Park et al. [67] investigated the diffusion behaviors in an extrusion process containing a mixing screw. It was observed that shear mixing caused by the screw rotation promotes convective diffusion. In convective diffusion, the screw motion creates the contact between a high gas concentration region (gas bubble) and a low gas concentration region (polymer melt). Furthermore, by stretching gas bubbles in the shear field generated by the motion of the screw which enhances the diffusion process, the interfacial area will be increased, thereby improving the diffusion mechanism. In addition, a dissolution enhancing device containing static mixers in the extrusion system will enhance the dissolution process by generating shear fields as the mixing elements are reorienting the melt along the flow direction, thus promoting solution formation.

#### 4.3.4.5 Cell Nucleation

Nucleation is a critical step in fine-celled or microcellular foaming processes. Nucleation can be defined as the transformation of small clusters of gas molecules into energetically stable groups or pockets. In order to create bubbles in liquids or polymer melts, a minimum amount of energy must be given to the system so that it can break the free energy barrier. This energy can be provided by heating or through a pressure drop. Two types of nucleation mechanisms can be observed: homogeneous and heterogeneous nucleation. Homogeneous nucleation is a type of nucleation where cells are nucleated randomly throughout the liquid or polymer melt matrix. It requires higher nucleation energy than heterogeneous nucleation. Heterogeneous nucleation is defined as nucleation at certain preferred sites, such as on the phase boundary, or sites provided by the additive particles.

Homogeneous Nucleation

In case of homogeneous nucleation, the classical nucleation theory has been used to describe the nucleation behavior in microcellular foaming by Colton and Suh. The nucleation rate is expressed as [68–74]:

$$N = C_0 f_0 \exp(-\frac{\Delta G^*}{kT}) \tag{4.7}$$

where $C_0$ is the concentration of gas molecules in the solution, $f_0$ is the frequency factor of gas molecules joining the nucleus, and $k$ is the Boltzman constant, $\Delta G^*_{hom}$ is the homogeneous critical nucleation energy required to form a nucleus with a critical size.

The critical nucleation energy is expressed as:

$$\Delta G^*_{hom} = \frac{16\pi\gamma^3_{pb}}{3\Delta P^2} \tag{4.8}$$

and the corresponding critical bubble size is:

$$r^* = \frac{2\gamma_{pb}}{\Delta P} \tag{4.9}$$

Here, $\gamma_{pb}$ is the liquid–gas surface tension, and $P$ is the pressure difference between that inside the critical nuclei and that around the surrounding liquid. Assuming that the polymer is fully saturated at by the blowing agent and the partial molar volume of blowing agent in the polymer is zero, $P$ can be taken as the saturation pressure [21, 68].

According to the classical nucleation theory, a greater number of cells can be nucleated as the saturation pressure, $\Delta P$, increases. The saturation pressure can be estimated as the gas concentration in the polymer according to Henry's law (Eq. 4.2). When the amount of gas in the polymer increases, the chance to nucleate more cells also increases. Even though the classical nucleation theory yields very valuable information about the pressure drop and cell nucleation relationship, it does not predict the effect of the pressure drop rate on cell nucleation. The effect of the pressure drop rate on nucleation is another important parameter and should be carefully examined. In the classical nucleation theory, an instantaneous pressure drop and instantaneous nucleation are assumed and thus, the nucleation rate, $N$ of Eq. (4.7) corresponds to the number of nucleated cells. However, in reality, the pressure drop is not instantaneous and happens over a finite time period. The nucleation rate will be affected based on how fast the pressure drops, or based on the pressure drop rate. The effect of the pressure drop rate on cell nucleation is as follows: the faster the pressure drops, the more that cells are nucleated [50]. Since the higher pressure drop rate requires a shorter time period; the already nucleated cells do not have a chance to grow a great deal. Therefore, more gas is utilized for cell nucleation and less is used for cell growth. As a result, high cell density foams or microcellular foams can be produced with high pressure drop rate dies.

Heterogeneous Nucleation

Heterogeneous nucleation is the other type of nucleation which is promoted at some preferred sites. With the existence of impurities such as nucleating agents, additives, or initiator residues, heterogeneous nucleation will occur. Typically, the nucleation takes place at the particle surface due to lowered activation energy.

The heterogeneous nucleation rate is expressed as follows [68–70]:

$$N_1 = C_1 f_1 \exp(-\frac{\Delta G^*_{\text{het}}}{kT}) \tag{4.10}$$

where $C_1$ the concentration of gas molecules, $f_1$ is the frequency factor of gas molecules joining the nucleus, $k$ is the Boltzman's constant, and $T$ is the temperature in K. $\Delta G^*_{\text{het}}$ is Gibbs free energy and can be expressed for heterogeneous nucleation, which occurs at smooth planar surfaces as follows (see Fig. 4.8a):

$$\Delta G^*_{\text{het}} = \frac{16\pi\gamma^3_{\text{pb}}}{3\Delta P^2} F(\theta_c) \tag{4.11}$$

where $\gamma^3_{\text{pb}}$ is the surface energy of the polymer–bubble interface, $\Delta P$ is the gas pressure used to diffuse the gas into the polymer. $F(\theta_c)$, which is the reduction of energy due to the inclusion of additives (nucleates), can be expressed as follows:

$$F(\theta_c) = \frac{1}{4}(2 + \cos\theta)(1 - \cos\theta)^2 \tag{4.12}$$

where $\theta_c$ is the contact angle of the polymer-additive gas interface.

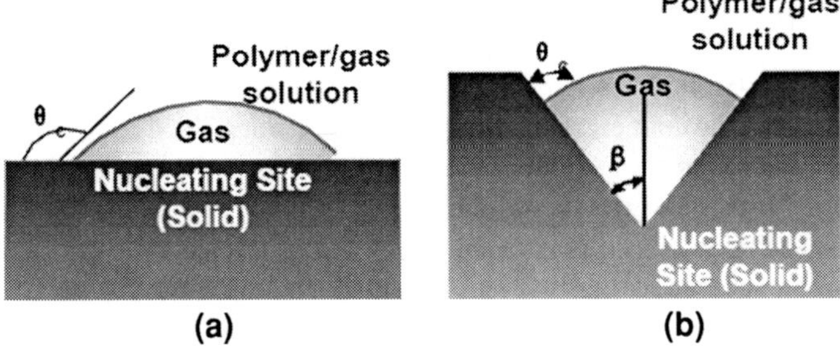


**Fig. 4.8** A bubble nucleated on **a** a smooth planar surface and **b** in a conical cavity

In actual polymeric foaming processes, the geometry of the nucleating sites, which depends on the nucleating agents themselves, the presence of unknown additives or impurities and the nature of the internal walls of equipment, varies from one site to another. Therefore, instead of assuming that all nucleating sites are either smooth planar surface, observable nucleation rates can occur in conical cavities that exhibit geometries consistent with the image presented in Fig. 4.8b where the semiconical angles, $\beta$ are randomly distributed between 0° and 90° at different nucleating sites. In this case, $F(\theta_c, \beta)$ is the reduction of energy, which can be expressed as [75]

$$F(\theta_c, \beta) = \frac{1}{4}\left[2 - 2\sin(\theta_c - \beta) + \frac{\cos\theta_c \cos^2(\theta_c - \beta)}{\sin\beta}\right] \tag{4.13}$$

Since the phase boundaries between the additive and the polymer matrix have a lower free energy barrier for nucleation than homogeneous nucleation, nucleation is more likely to occur at these sites. The number of cells nucleated can be controlled by the amount of additives [76, 77]. Also, if the additive particle size is fine graded (less than a micron) and well dispersed in the polymer matrix, a uniformly distributed microcell structure can be produced [76]. Lee et al. [78] examined the gas absorption behavior of polymer systems to explain heterogeneous nucleation in mineral filled polymers: HDPE with/without talc, and PVC with/without $CaCO_3$. It was suggested that the accumulated gas in the filler–polymer interface helps to create cells in the foaming process. Ramesh et al. [79] developed a model for heterogeneous nucleation in the blend of PS and high impact polystyrene (HIPS) based on the presence of microvoids.

#### 4.3.4.6 Cell Growth

After cells are nucleated, they continue to grow because of gas diffusion from the polymer matrix. Since the pressure inside the cells is greater than the surrounding pressure, cells tend to grow in order to decrease the pressure difference between the inside and the outside [20]. The cell growth mechanism is affected by the viscosity, diffusion coefficient, gas concentration, and the number of nucleated cells. The temperature can control the amount of cell growth, which then affects two important parameters: diffusivity and melt viscosity. For instance, if the temperature decreases, the diffusivity of gas decreases and the melt viscosity of the matrix increase, thus, decreasing the cell growth rate. In the foaming process, maintaining the gas in the polymer matrix by close temperature control is essential for achieving good cell growth and thus, high volume expansion. In microcellular foams, because the cell size is very small and the cell density is very high, the cell wall thickness separating the two cells is smaller and the rate of growth is faster than in conventional foams.

Cell growth is a process that involves mass, momentum and heat transportations of the fluid. Almost all the models describing the cell growth evolve from the "cell model", a model used to describe the cell growth from a single bubble that is surrounded by an infinite sea of fluid with an infinite amount of available gas. Each cell is assumed to have equal and constant mass with a spherical structure. A schematic diagram of the bubble growth is shown in Fig. 4.9 [80]. With several assumptions and simplifications, the equation of motion, the integral mass (gas) balance over the bubble, and the differential mass (dissolved gas) balance in the surrounding liquid phase take the following forms:

$$\frac{dR_{\mathrm{b}}}{dt} - \frac{(P_{\mathrm{G}} - P_{\mathrm{L}})R_{\mathrm{b}}}{4\eta} = \frac{\gamma_{\mathrm{lv}}}{2\eta} \tag{4.14}$$

$$\frac{d}{dt}\left(\frac{4\pi P_{\mathrm{G}} R_{\mathrm{b}}^{3}}{3\Re T}\right) = 4\pi R_{\mathrm{b}}^{2} D \frac{\partial c}{\partial r}\bigg|_{r=R} \tag{4.15}$$

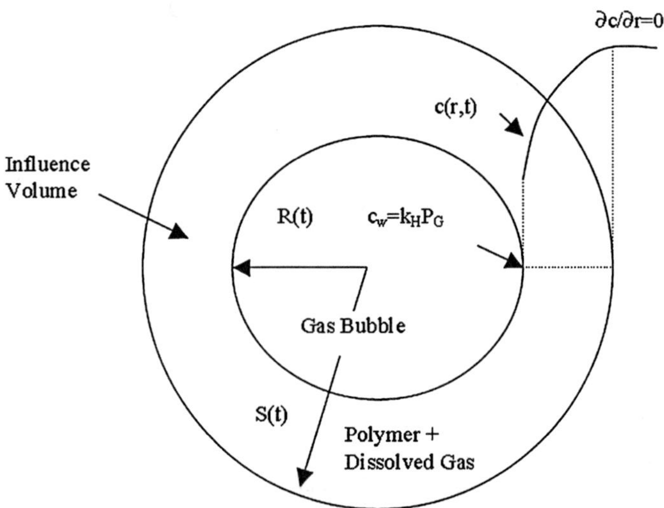


**Fig. 4.9** Schematic diagram of bubble growth in a liquid shell

$$\frac{\partial c}{\partial t} + \frac{\dot{R}_b R^2 \partial c}{r^2 \ \partial r} = \frac{D \partial}{r^2 \partial t}\left(r^2 \frac{\partial c}{\partial r}\right) \tag{4.16}$$

where $P_G$ and $P_L$ are the pressure in the bubble and in the polymer melt, respectively. $R_b$ is the radius of the bubble, $\dot{R}_b$ is the rate of bubble growth, $\eta$ is the Newtonian viscosity, $\Re$ is the gas constant, $c$ is the gas concentration, and $D$ is diffusivity.

However, this may also cause cell coalescence, which is undesirable [81]. If the cells coalesce during cell growth, the initial cell density will be deteriorated. As nucleated cells grow, adjacent cells will begin to touch each other. These contiguous cells tend to coalesce because the total free energy is lowered by reducing the surface area of cells via cell coalescence [20]. It may be noted that the shear field generated during the shaping process tends to stretch nucleated bubbles, which further accelerates cell coalescence [81]. When the cell density is deteriorated, the mechanical and thermal properties are deteriorated as well.

Although Baldwin et al. [82] attempted to prevent cell coalescence in the die by pressurizing the nucleated polymer solution during shaping, the extruded foam structure showed that many adjacent cells were coalesced and that the cell density was deteriorated. Considering the difficulty of maintaining a high backpressure in the shaping die in case of a large cross section of extruded foam, it may not be realistic to prevent cell coalescence by controlling the pressure alone in the shaping die. Park et al. [81] suggested a means for suppressing cell coalescence by increasing the melt strength of polymer via temperature control in microcellular extrusion processing. The melt strength, by definition, may be treated as a degree of resistance to the extensional flow of the cell wall during the drainage of polymer in the cell wall when volume expansion takes place. Therefore, the cell wall stability will increase as the melt strength increases [26].

### 4.3.5 Mechanical Properties

Although a wide range of polymers have been successfully synthesized into microcellular parts such as PS, PVC, PC, PP and PMMA, reports on their mechanical properties are still limited.

The status of previous research on the mechanical properties of microcellular foams can be summarized as follows. For most polymer systems, microcellular foams exhibited superior impact strength, toughness and fatigue life compared to their solid counterparts. However, the extent of improvement differs among different systems. Further, even for the same polymer-gas system, discrepancy in results was reported by different research groups. Finally, direct comparison of the mechanical properties between microcellular foams and their macrocelluar counterparts with the same density is extremely limited. A lot of research has been done which focused on the impact, tensile, compressive, and fatigue properties [83–91].

### *4.3.6 Thermoplastic Elastomers Foams*

Thermoplastic elastomers (TPEs) are increasingly attracting attention as an alternative class of materials to thermosetting rubbers. The most attractive feature of these materials is that they can be processed like thermoplastics while exhibiting the resilience and elasticity characteristic of elastomers. The great majority of TPEs have hetero-phase morphology, whether the TPE is derived from block copolymers, rubber–plastic compositions or ionomers. In general, the hard domains undergo molecular relaxation at elevated temperatures, thus allowing the material to flow. When cooled, the hard domains (i.e., the thermoplastic regions) again solidify and provide tensile strength at normal service temperatures [92, 93]. The soft domains give the material the elastomeric characteristics.

The market of TPEs based on polyolefin rubber–plastic compositions has grown along two distinctly different product lines: one class consists of a simple blend and classically meets the definition of a Thermoplastic Olefins (TPO), and in the other class the rubber phase is dynamically vulcanized giving rise to thermoplastic vulcanizate (TPV). Both the simple blend and the dynamically vulcanized TPE have found wide industrial applications [94]. TPV is prepared by melt mixing a thermoplastic with an elastomer in the presence of a small quantity of vulcanizing system which leads to the in situ crosslinking of the rubber phase. TPV was first introduced by Gessler et al. [95] and since then has attracted great industrial attention [96]. Currently, TPV is replacing EPDM rubber dramatically because of the impressive advantages for automotive sealing applications. Some of the advantages of TPV compared to that of EPDM rubber are good gloss, recyclability, improved colorability, and shorter cycle time and design flexibility.

The development of TPV foaming technology is to fulfill the requirement of achieving lower cost, lighter weight and better fuel economy. Foaming of TPV has not been investigated extensively. The first extensive investigation of foaming in TPV was carried out in 1992 by Dutta and Cakmak [97]. They studied foaming of an olefinic thermoplastic elastomer; consisting of polypropylene (PP)/ethylene-propylenediene-terpolymer (EPDM), commercially known as Santoprene thermoplastic elastomer, using a CBA that releases $N_2$ upon decomposition. However, the foam densities they were able to achieve were limited to 0.5 g/cm$^3$. The foam structure was non-homogeneous and a wide cell size distribution was obtained.

Later efforts have been taken by one of the major TPV suppliers, DSM of Netherlands, to develop a foaming technology which involves a water releasing chemical compound (WCC) as a CBA [98, 99]. DSM also introduces a foaming grade TPV, Sarlink. DSM claims that their foaming technology can be used to produce TPV foams with a density from 0.15 to 0.90 g/cm$^3$. The advantage of a CBA is that it doesn't require a special foaming extrusion system but CBA are usually expensive so an additional 10–15% cost will be added up to the final TPV foam product. In addition, due to unexpected reactions during decomposition of CBA and foaming, CBA-blown TPV foam usually has an open-celled structure which leads to more water absorption that is not favorable for the sealing purpose.

Another commercialized TPV foaming technology is based on water as a PBA which has been independently developed by Sahnoune from AES [100]. Water is an attractive PBA from the standpoint of cost and environmental requirements. It also does not require special handling, is readily available, and the pressures involved in the foaming process are not as demanding as for other blowing agents. Sahnoune et al. [100] studied the effect of water concentration on foam density and cell nucleation under the various processing conditions for a TPV. His study found that, despite of their complex microscopic structure, TPV foam very much like conventional thermoplastics do. The water foaming method can achieve from density as low as 0.15 g/cm$^3$ but the cell size is around 200–300 μm.

Spitael et al. [101] studied the foaming behaviors of several TPV formulations containing various amounts of branched polypropylene resin with water as the blowing agent, while the extensional viscosity of the materials with different formulations was measured and considered. They indicated that the replacement of a small amount of linear polypropylene with branched polypropylene improve the foam density and cellular structure. However, as the added content of branched polypropylene was increased, a worse foamability was observed. They concluded that there exists an optimal amount of branched content.

More recently, Kropp et al. [102] carried out a comparative study of foaming three types of TPEs with carbon dioxide ($CO_2$) and using hydrocerol as a nucleating agent. These materials were thermoplastic polyurethane (TPU), a styrene based TPE (SEBS) and again a PP/EPDM TPV. They found that foaming was most difficult with the PP/EPDM TPV. The foam density they could reach was 0.76 g/cm$^3$ and the foam structure was also not uniform. They found that a specific TPU type showed the best foamability; a SEBS-type was also successfully foamed; and a PP/EPDM TPV especially developed for water foaming was most difficult to foam with $CO_2$ as the blowing agent. Kim et al. [103, 104] extensively studied the foamability of commercial grade thermoplastic vulcanizates (TPVs) with carbon dioxide and nitrogen in the extrusion system, the results showed that TPV foam with $N_2$ produced a uniform and fine cell structure with a smooth surface, and indicated that $N_2$ could be a very good PBA for fine cell TPV foam. Wong et al. [59] reported the microcellular TPO foams with a two-stage batch process method using $N_2$ as blowing agent.

All the foaming of polyolefin elastomer blends showed that there is no different than foaming of any other polymer, although which has a more complex microscopic structure and is multiphase systems made of at least two phases: the thermoplastic phase and the rubber phase.

## References

1. Arbelaiz, A., Fernández, B., Cantero, G., Llano-Ponte, R., Valea, A., Mondragon, I.: Mechanical properties of flax fibre/polypropylene composites: influence of fibre/matrix modification and glass fibre hybridization. Compos. Part A **36**, 1637–1644 (2005)

2. Cantero, G., Arbelaiz, A., Llano-Ponte, R., Mondragon, I.: Effects of fibre treatment on wettability and mechanical behaviour of flax/polypropylene composites. Compos. Sci. Technol. **63**, 1247–1254 (2003)
3. Raj, R.G., Kokta, B.V., Groleau, G., Daneault, C.: Use of wood fiber as a filler in polyethylene: studies on mechanical properties. Plast. Rubber Process. Appl. **11**(4), 215 (1989)
4. Bledzki, A.K., Faruk, O.V., Sperber, E.: Cars from bio-fibres. Macromol. Mater. Eng. **291**, 449–457 (2006)
5. Clemons, C.: Wood-plastic composites in the United States: the interfacing of two industries. For. Prod. J. **52**(6), 10–18 (2002)
6. Kokta, B.V., Maldas, D., Daneault, C., Beland, P.: Composites of polyvinyl chloride-wood fibers: I. Effect of isocyanate as a bonding agent. Polym. Plast. Technol. Eng. **29**(1/2), 87–118 (1990)
7. Pickering, K.L., Abdalla, A., Ji, C., McDonald, A.G., Franich, R.A.: The effect of silane coupling agents on radiata pine fibre for use in thermoplastic matrix composites. Compos. Part A. Appl. Sci. Manuf. **34**(10), 915–926 (2003)
8. Kazayawoko, M., Balatinecz, J.J., Matuana, L.M.: Surface modification and adhesion mechanisms in wood fiber-polypropylene composites. J. Mater. Sci. **34**, 6189–6199 (1999)
9. Coutinho, F.M.B., Costa, T.H.S., Carvalho, D.L.: Polypropylene-wood fiber composites: effect of treatment and mixing conditions on mechanical properties. J. Appl. Polym. Sci. **65**, 1227–1235 (1997)
10. Raj, R.G., Kokta, B.V., Maldas, D., Daneault, C.: Use of wood fibers in thermoplastics. VII: The effect of coupling agents in polyethylene-wood fibers composites. J. Appl. Polym. Sci. **37**, 1089–1103 (1989)
11. Saheb, D.N., Jog, J.P.: Natural fiber polymer composites: a review. Adv. Polym. Technol. **18**(4), 351–363 (1999)
12. Pape, P.G., Romenesko, D.J.: The role of silicone powders in reducing the heat release rate and evolution of smoke in flame retardant thermoplastics. J. Vinyl Addit. Technol. **3**, 225–232 (1997)
13. Le Bras, M., Wilkie, C.A., Bourbigot, S., Duquesne S., Jama, C.: Fire Retardancy of Polymers: New Applications of Mineral Fillers, chap. 2. The Royal Society of Chemistry Publisher, UK (2005)
14. Wei, P., Hao, J., Du, J., Han, Z., Wang, J.: An investigation on synergism of an intumescent flame retardant based on silica and alumina. J. Fire Sci. **21**, 17–28 (2003)
15. Gilman, J.W., Kashiwagi, T., Harris, R.H., Jr., Lomakin, S., Lichetenhan, J.D., Jones, P., Bolf, A.: In: Al-Malaika, S., Wilkie, C., Golovoy, C.A. (eds.) Chemistry and Technology of Polymer Additives. Blackwell Science, London (1999)
16. Kashiwagi, T., Gilman, J.W., Butler, K.M., Harris, R.H., Shields, J.R., Asano, A.: Flame retardant mechanism of silica gel/silica. Fire Mater. **24**, 277–289 (2000)
17. Sain, M., Park, S.H., Suhara, F., Law, S.: Flame retardant and mechanical properties of natural fiber-PP composites containing magnesium hydroxide. Polym. Degrad. Stab. **83**(2), 363–367 (2004)
18. Zhao, Y., Wang, K., Zhu, F., Xue, P., Jia, M.: Properties of poly(vinyl chloride)/wood flour/montmorillonite composites: effects of coupling agents and layered silicate. Polym. Degrad. Stab. **91**(2), 2874–2883 (2006)
19. Li, B., He, J.: Investigation the mechanical property, flame retardancy and thermal degradation of LLDPE-wood fiber composites. Polym. Degrad. Stab. **83**, 241–246 (2004)
20. Klempner, D., Frisch, K.C.: Handbook of Polymeric Foams and Foam Technology. Oxford University, New York (1991)
21. Landrock, A.H.: Handbook of Plastic Foams: Types, Properties, Manufacture and Applications. Noyes, NJ (1991)
22. Lee, S.T.: Foam Extrusion: Principles and Practice. CRC Press, London (2000)
23. Baker, R.W.: Membrane Technology and Application. McGraw Hill, New York (2000)
24. Hedrick, J.L., Carter, K.R., Labadie, J.W.: Nanoporous polyimides. Adv. Polym. Sci. **141**, 1–8 (1999)

25. Luebke, G., Holzberg, T.: New developments of chemical foaming agents for wood plastic composites. In: Fourth International Wood and Natural Fibre Composites Symposium, Kassel, Germany, p. 15-1 (2002)
26. Frich, K.C., Saunders, J. H.: Plastic Foams, Part I. Marcel Dekker Inc., New York (1972)
27. Gorski, R.A., Ramsey, R.B., Dishart, K.T.: Physical properties of blowing agent polymer systems-I: solubility of fluorocarbon blowing agents in thermoplastic resins. J. Cell. Plast. **22**, 21–52 (1986)
28. Dwyer, F.J., Zwolinski, L.M., Thrun, K.M.: Extruding thermoplastic foams with a non-CFC blowing agent. Plast. Eng. **5**, 29–32 (1990)
29. Mccallum, T.J.: Properties and foaming behaviour of thermoplastic olefin blends based on linear and branched polypropylene. PhD Dissertation, Queen's University, p. 21 (2007)
30. Shutov, F.A.: Integral/structureal polymer foams. Springer, New York (1986)
31. Botillier, P.E.: Br Patent 1184688 (1969)
32. Martini, J., Waldrnan, F.A., Suh, N.P.: The production and analysis of microcellular thermoplastic foams. SPE ANTEC Technical Papers, vol. 28, p. 674 (1982)
33. Doroudiani, S., Park, C.B., Kortschot, M.T.: Processing and characterization of microcellular foamed high-density polyethylene/isotactic polypropylene blends. Polym. Eng. Sci. **38**(7), 1205–1215 (1998)
34. Kumar, V., Weller, J.E.: A process to produce microcellular PVC. Int. Polym. Process. **VIII**(1),73–80 (1993)
35. Kumar, V., Weller, J.: Production of microcellular polycarbonate using carbon dioxide for bubble nucleation. J. Eng. Ind. **116**, 413–420 (1994)
36. Kumar, V., Schirmer, H.G.: Semi-continuous production of solid state polymeric foams. US Patent 5,684,055 (1997)
37. Kumar, V., Schirmer, H.G.: Semi-continuous production of solid-state PET foams. SPE-ANTEC, vol. 2, pp. 2189–2192 (1995)
38. Rabinovitch, E.B., Isner, J.D., Sidor, J.A., Wiedl, D.J.: Effect of extrusion conditions on rigid PVC foam. J. Vinyl Addit. Technol. **3**, 210–213 (1997)
39. Mengeloglu, F., Matuana, L.M.: Foaming of rigid PVC/wood-flour composites through a continuous extrusion process. J. Vinyl Addit. Technol. **7**, 142–148 (2001)
40. Mengeloglu, F., Matuana, L.M.: Mechanical properties of extrusion-foamed rigid PVC/ wood-flour composites. J. Vinyl Addit. Technol. **9**, 26–31 (2003)
41. Lee, S.T., Kareko, L., Jun, J.: Study of thermoplastic PLA foam extrusion. J. Cell. Plast. **44**, 293–305 (2008)
42. Jeong, B., Xanthos, M., Seo, Y.: Extrusion foaming behavior of PBT resins. J. Cell. Plast. **42**, 165–176 (2006)
43. Zhang, S., Rodrigue, D.: Preparation and morphology of polypropylene/wood flour composite foams via extrusion. Polym. Compos. **26**, 731–738 (2005)
44. Li, Q., Matuana, L.M.: Foam extrusion of high density polyethylene/wood-flour composites using chemical foaming agents. J. Appl. Polym. Sci. **88**, 3139–3150 (2002)
45. Han, X., Koelling, K.W., Tomasko, D.L., Lee, L.J.: Continuous microcellular polystyrene foam extrusion with supercritical $CO_2$. Polym. Eng. Sci. **42**(11), 2094–2106 (2004)
46. Han, X., Zeng, C., Lee, L.J., Koelling, K.W., Tomasko, D.L.: Extrusion of polystyrene nanocomposite foams with supercritical $CO_2$. Polym. Eng. Sci. **43**(6), 1261–1275 (2004)
47. Lee, M., Tzoganakis, C., Park, C.B.: Extrusion of PE/PS blends with supercritical carbon dioxide. Polym. Eng. Sci. **38**(7), 1112–1120 (2004)
48. Siripurapu, S.S., Gay, Y.J., Royer, J.R., Desimone, J.M., Spontak, R.J.: Generation of microcellular foams of PVDF and its blends using supercritical carbon dioxide in a continuous process. Polymer **43**(20), 5511–5520 (2002)
49. Park, C.B., Suh, N.P.: Filamentary extrusion of microcellular polymers using a rapid decompressive element. Polym. Eng. Sci. **36**(1), 34–48 (1996)
50. Park, C.B., Baldwin, D.F., Suh, N.P.: Effect of the pressure drop rate on cell nucleation in continuous processing of microcellular polymers. Polym. Eng. Sci. **35**(5), 432–440 (1995)

51. Park, C.B., Suh, N.P.: Rapid polymer/gas solution formation for continuous production of microcellular plastics. J. Manuf. Sci. Eng. **118**, 639–645 (1996)
52. Guo, M.C., Heuzey, M.C., Carreau, P.J.: Cell structure and dynamic properties of injection molded polypropylene foams. Polym. Eng. Sci. **47**, 1070–1081 (2007)
53. Martini-Vvedensky, J.E., Suh, N.P., Waldman, F.A.: Microcellular closed cell foams and their method of manufacture. US Patent 4,473,665 (1984)
54. Trexel. http://www.trexel.com
55. Xu, J.: Reciprocating-screw injection molding machine for microcellular foam. SPE-ANTEC, pp. 449–453 (2001)
56. Jacobsen, K., Pierick, D.: Microcellular foam molding: advantages and application examples. SPE-ANTEC, pp. 1929–1933 (2000)
57. Moore, S.: Foam molding resurgence: sparks competition among processes. Mod. Plast. **11**, 23–25 (2001)
58. Kishbaugh, L.A., Levesque, K.J., Guillemette, A.H., Chen, L., Xu, J., Okamoto, K.T.: Fiber-filled molded foam articles, molding, and process aids (USA). WO:2002026482 (2002)
59. Wong, S., Lee, J.W.S., Naguib, H.E., Park, C.B.: Effect of processing parameters on the mechanical properties of injection molded thermoplastic polyolefin (TPO) cellular foams. Macromol. Mater. Eng. **293**, 605–613
60. Crank, J., Park, G.S.: Diffusion in polymers. Academic Press Inc., New York (1968)
61. Van Krevelen, D.W.: Properties of polymers. Elsevier, New York (1990)
62. Wissinger, R.G., Paulaitis, M.E.: Swelling and sorption in polymer-$CO_2$ mixtures at elevated pressures. J. Polym. Sci. Part B. Polym. Phys. **25**, 2497–2510 (1987)
63. Wissinger, R.G., Paulaitis, M.E.: Molecular thermodynamic model for sorption and swelling in glassy polymer-$CO_2$ system at elevated pressures. Ind. Eng. Chem. Res. **2530**, 842–851 (1991)
64. Sato, Y., Fujiwara, K., Takikawa, T., Sumarno, Takishima, S., Masuoka, H.: Solubilities and diffusion coefficients of carbon dioxide and nitrogen in polypropylene, high-density polyethylene, and polystyrene under high pressures and temperatures. Fluid Phase Equilib. **162**, 261–276 (1999)
65. Sato, Y., Takikawa, T., Takishima, S., Masuoka, H.: Solubility and diffusion coefficient of carbon dioxide in poly(vinyl acetate) and polystyrene. J. Supercrit. Fluids **19**, 187–198 (2001)
66. Zhang, Q., Xanthos, M., Dey, S.K.: In-line measurement of gas solubility in polystyrene and polyethylene terephthalate melts during foam extrusion. MD (Am. Soc. Mech. Eng.), vol. 82 (Porous, Cellular and Microcellular Materials), pp. 75–83 (1998)
67. Park, C.B., Suh, N.P.: Rapid polymer/gas solution formation for continuous processing of microcellular plastics ASME trans. J. Manuf. Sci. Eng. **118**, 639–645 (1996)
68. Colton, J.S., Suh, N.P.: Nucleation of microcellular foam: theory and practice. Polym. Eng. Sci. **27**, 500–503 (1987)
69. Colton, J.S., Suh, N.P.: Nucleation of microcellular thermoplastic foam with additives: part 1. Theoretical considerations. Polym. Eng. Sci. **27**(7), 485–492 (1987)
70. Colton, J.S., Suh, N.P.: Nucleation of microcellular thermoplastic foam with additives: part 2. Experimental results and discussion. Polym. Eng. Sci. **27**(7), 493–499 (1987)
71. Lee, J.H., Flumerfelt, R.W.: A refined approach to bubble nucleation and polymer foaming process: dissolved gas and cluster size effects. J. Coll. Interface Sci. **184**, 335–348 (1996)
72. Punnathanam, S., Corti, D.S.: Homogeneous nucleation in stretched fluids: cavity formation in the superheated Lennard-Jones liquid. Ind. Eng. Chem. Res. **41**, 1113–1121 (2002)
73. Han, J.H., Han, C.D.: A study on bubble nucleation in polymeric liquid. I. bubble nucleation in concentrated polymer solutions. J. Polym. Sci. Part B. Polym. Phys. **28**, 711–741 (1990)
74. Han, J.H., Han, C.D.: A study on bubble nucleation in polymeric liquid. II. Theoretical consideration. J. Polym. Sci. Part B. Polym. Phys. **28**, 743–761 (1990)
75. Leung, S.N., Park, C.B, Li, H.: Numerical simulation of polymeric foaming processes using modified nucleation theory. Plast. Rubber Compos. **35**, 93 (2006)

76. Colton, J.S., Suh, N.P.: The nucleation of microcellular thermoplastic foam with additives. Part II: experimental results and discussion. Polym. Eng. Sci. **27**, 493–499 (1987)
77. Behravesh, A.H., Park, C.B., Cheung, L.K., Venter, R.D.: Extrusion of polypropylene foams with hydrocerol and isopentane. J. Vinyl Addit. Technol. **2**(4), 349–357 (1996)
78. Lee, C., Sheth, S.H., Kim, R.: Gas absorption with filled polymer systems. Polym. Eng. Sci. **41**(6), 990–997 (2001)
79. Ramesh, N.S., Rasmussen, D.H., Campbell, G.A.: The heterogeneous nucleation of microcellular foams assisted by the survival of microvoids in polymers containing low glass-transition particles. 1. Mathematical-modeling and numerical simulation. Polym. Eng. Sci. **34**, 1685–1697 (1994)
80. Lee, S.T., Ramesh, N.S.: In: Kumar, V., Seeler, K.A. (eds.) Cellular and Microcellular Materials, **76**, pp. 71–80. ASME, New York (1996)
81. Park, C.B., Behravesh, A.H., Venter, R.D.: Chapter 8: a strategy for suppression of cell coalescence in the extrusion of microcellular HIPS foams. In: Khemani, K. (ed.) Foam book: recent advances in polymeric foam science and technology. ACS, Washington, pp. 115–129 (1997)
82. Baldwin, D.F., Park, C.B., Suh, N.P.: A microcellular processing study of poly(ethylene terephtalate) in the amorphous and semicrystalline states: part II. Cell growth and process design. Polym. Eng. Sci. **36**, 1446–1453 (1996)
83. Barlow, C., Kumar, V., Flinn, B., Bordia, R.K., Weller, J.: Impact strength of high density solid-state microcellular polycarbonate foams. J. Eng. Mater. Technol. **123**(2), 229–233 (2001)
84. Collias, D.I., Baird, D.G., Borggreve, R.J.M.: Impact toughening of polycarbonate by microcellular foaming. Polymer **35**(18), 3978–3983 (1994)
85. Collias, D.I., Baird, D.G.: Tensile toughness of microcellular foams of polystyrene, styrene-acrylonitrile copolymer, and polycarbonate, and the effect of dissolved gas on the tensile toughness of the same polymer matrixes and microcellular foams. Polym. Eng. Sci. **35**(14), 1167–1177 (1995)
86. Matuana, L.M., Park, C.B., Balatinecz, J.J.: Structures and mechanical properties of microcellular foamed polyvinyl chloride. Cell. Polym. **17**(1), 1–16 (1998)
87. Kumar, V.: Microcellular plastics: does microcellular structure always lead to an improvement in impact properties? 60th SPE-ANTEC, vol. 2, pp. 1892–1896 (2002)
88. Juntunen, R.P., Kumar, V., Weller, J.E., Bezubic, W.P.: Impact strength of high density microcellular poly(vinyl chloride) foams. J. Vinyl Addit. Technol. **6**(2), 93–99 (2000)
89. Seeler, K.A., Kumar, V.: Tension-tension fatigue of microcellular polycarbonate: initial results. J. Reinf. Plast. Compos. **12**(3), 359–376 (1993)
90. Kumar, V., VanderWel, M., Weller, J., Seeler, K.A.: Experimental characterization of the tensile behavior of microcellular polycarbonate foams. J. Eng. Mater. Technol. **116**(4), 439–445 (1994)
91. Arora, K.A, Lesser, A.J, McCarthy, T.J.: Compressive behavior of microcellular polystyrene foams processed in supercritical carbon dioxide. Polym. Eng. Sci. **38**(12), 2055–2062 (1998)
92. Legge, N.R., Holden, G., Schroeder, H.E.: Thermoplastic elastomer: a comprehensive review. Hanser Publishers, Munich (1987)
93. Walker, B.M., Rader, C.P.: Handbook of Thermoplastic Elastomers. Van Nostrand Reinhold Co., New York (1998)
94. Abdou-Sabet, S., Patel, R.P.: Morphology of elastomeric alloys. Rubber Chem. Technol. **64**, 769–779 (1991)
95. Gessler, A.M., Haslett, W.H.: Process for preparing a vulcanized blend of crystalline polypropylene and chlorinated butyl rubber. US Patent 3,037,954 (1962)
96. Gottler, W.K., Richwine, J.R., Wille, F.J.: The rheology and processing of olefin-based thermoplastic vulcanizates. Rubber Chem. Technol. **55**, 1448–1463 (1982)
97. Dutta, A., Cakmak, M.: Influence of composition and processing history on the cellular morphology of the foamed olefinic thermoplastic elastomers. Rubber Chem. Technol. **65**, 932–955 (1992)

98. Brzoskowski, R., Wang, Y., La Tulippe, C., Dion, B., Cai, H., Sadeghi, H.: Extrusion of low density chemically foamed thermoplastic vulcanizates. SPE-ANTEC (Annual Technical Conference) Technical Papers, vol. 3, pp. 3204–3208 (1998)
99. Wang, Y., Cai, H., Freitas, L., Dion, B., Brzoskowski, R.: TPV foaming with water-releasing compound. Kunststoffe Plast. Eur. **88**(12), 2170–2172 (1998)
100. Sahnoune, A.: Foaming of thermoplastic elastomers with water. J. Cell. Plast. **37**(2), 149–159 (2001)
101. Spitael, P., Macosko, C.W.: Strain hardening in polypropylenes and its role in extrusion foaming. Polym. Eng. Sci. **44**(11), 2090–2100 (2004)
102. Kropp, D., Michaeli, W., Herrmann, T., Schroder, O.: Foam extrusion of thermoplastic elastomers using $CO_2$ as blowing agent. J. Cell. Plast. **34**(4), 304–311 (1998)
103. Kim, S.G., Park, C.B., Kang, B.S., Sain, M.: Foamability of thermoplastic vulcanizates (TPVs) with carbon dioxide and nitrogen. Cell. Polym. **25**, 19–33 (2006)
104. Kim, S.G., Park, C.B., Kang, B.S., Sain, M.: Foamability of thermoplastic vulcanizates blown with various physical blowing agents. J. Cell. Plast. **44**, 53–67 (2008)

# Chapter 5
# Effect of Compatibilizers in WPC Composites

## 5.1 Introduction

Wood–fiber (WF) filled plastic composites (WPC) have gained rapid growth in recent years. Such materials offer significant advantages, which justify their use. Wood fiber is obtained from natural resources, it is available in various forms in large quantities, light, cheap, and it can be added to commodity matrices in considerable amounts thus offering economically advantageous solutions [1–4]. The main drawbacks of such composites are their poor adhesion to basically all matrix polymers [5, 6] and high density compared to natural wood and certain plastics [7].

Improvement in the mechanical properties of such composites requires strong adhesion between the wood fiber and polymer matrix. However, both components are incompatible, due to WF's hydrophilic and the polymer hydrophobic nature. In many studies, the compatibility between composite components was improved using either physical or chemical modification of the polymer or WF or by using coupling agents. Such as maleated polypropylenes (PP-g-MAs) and SEBS-g-MA have been known to increase adhesion between WF and polyolefin resins, resulting in improvements in the physical properties of WF/PP composites [8–12]. The final properties of such composites to a large extent depend on the compounding and processing conditions. Effective mixing is crucial for achieving optimal dispersion of WF and optimizes the properties of the composites.

In terms of processing, Yam et al. [13] suggested that mechanical properties and fiber length are sensitive to extrusion parameters such as screw configuration and compounding temperature. Twin screw extruders have come to play an important role in blending and compounding of polymers. Significant applications in the plastics industry started in the 1930s and 40s with the efforts of Colombo, Pasquetti, Maskat, Erdmenger, Leistritz and Fuller [14].

Among the processing equipment of WPC, both the counter-rotating and the co-rotating twin-screw extruder (TSE) are widely used, especially the intermeshing,

J. K. Kim and K. Pal, *Recent Advances in the Processing of Wood–Plastic Composites*, Engineering Materials, 32, DOI: 10.1007/978-3-642-14877-4_5,

co-rotating type. Fully intermeshing twin screws provide a narrow residence time distribution; therefore, provide uniform heat history to most wood fibers, which prevents wood degradation. Co-rotating screws are effective in alternating the direction of applied stresses through the use of different mixing elements, thus producing different mixing effects. Two types of mixing are widely used (i.e., the dispersive mixing and the distributive mixing) [15]. Dispersive mixing is the breaking up of clumps or agglomerates of particles into the ultimate particulate size. Distributive mixing is the distribution of particles without decreasing their size.

Recently, screw designs and configurations have attracted considerable attention in the compounding field with the appearance of modular type twin screw extruders [16–19]. Schmidt et al. [20] described an effort to develop the mixing model in a modular intermeshing twin screw extruders. Later, Kim and White [13] developed the distributive mixing model in a modular machine based on the flow model of Lyu et al. [21]. Kim and his coworkers [22] reported the flow behavior of waste EPDM and PP polymer using five different screw configurations in a co-rotating twin screw extruder and also developed a computer simulation program which showed the pressure distribution, temperature profile and fill factor on the screw configuration. Zhang et al. [23] investigated the screw configuration and screw speed on the dispersion of wood fiber in high density polyethylene. The results showed the best uniformity by medium dispersive and distributive mixing. However, limited studies demonstrate the effect of machine characteristics on reactive mixing of WPC and effect on foamability.

Although wood fiber reinforced polymer composites have been commercialized, their potential for use in many industrial applications has been limited because of their brittleness, lower impact resistance, and mainly higher density compared to neat plastics. The concept of creating microcellular foamed structures in the composites as a means to improve these shortcomings has successfully been demonstrated. A pressure-quench method described by Goel and Beckman [24] is widely used for making microcellular polymers via supercritical carbon dioxide ($scCO_2$). They found that the microcellular structure could be achieved by rapid depressurization to allow the cells nucleation and growth as in the batch process after saturating polymers with $scCO_2$. Recent studies have focused on the production of microcellular foams with high cell density by using nano-scaled nucleating agents [25–28], of which organically modified layered silicates were the most commonly used [28, 29], The homogeneous nanoparticle dispersion, favorable surface property and particle geometry account for the significant increase in cell density and decrease in cell diameter by adding a small amount of nanoparticles [26]. The microcellular foaming process used in this study was performed in our lab as pressure-quench batch process.

Some researchers in our lab have done this type of research. In this investigation, PP/WF composite are produced using various modular screw configurations in a co-rotating twin screw extruder and the corresponding variation of the mechanical properties, foaming properties and morphology are studied. The optimized screw configuration in then used in further detailed study of the effect of screw speed, fumed silica and compatibilizers effect on the mechanical properties, foaming properties and morphology are also investigated.

## 5.2 Preparation PP/Wood–Fiber Composites by Twin Screw Extruder

All experiments were carried out on a Bau Tech BA-19 twin screw extruder. It has a screw diameter of 19 mm and the distance between screw axes is 18.4 mm. Three different screw configurations were used in this study designated as A, B, and C. Each screw is an assembly of different numbers of right-handed and left-handed screw elements with right and left handed and neutral kneading blocks positioned in Fig. 5.1a–c. These screw assemblies were labeled as screws configuration A, B, and C. Screw C has one sets of special screw element which having a reverse-pumping effect. It was placed in the third kneading disc block, which increasing the residence time during processing. The detailed screw configurations of these four screws are shown in Fig. 5.1a–c.

### *5.2.1 Blending Process*

Polypropylene (R520Y), wood–fiber and compatibilizers were mixed and fed at a rate of 0.5 kg per hour through the hopper using a loss-in-weight feeder into the twin screw extruder. The screw speed studied was 50–200 rpm range and temperatures profile of the extruder from feeding to the die was 140°/160°/170°/170°/175°/175°/175°/180°C for all blends. To study the effects of processing parameters, different screw configurations, screw speeds, and compatibilizer were evaluated.

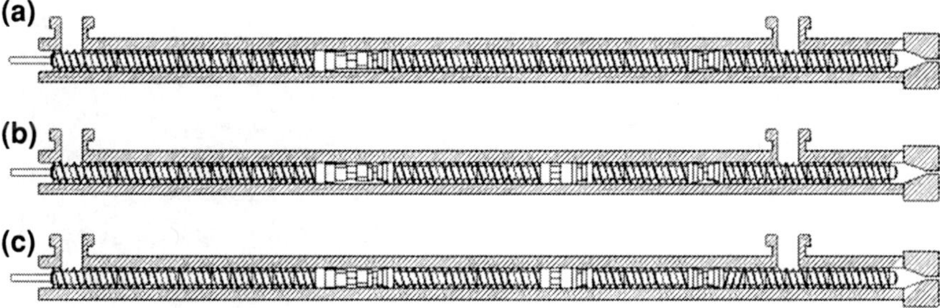


**Fig. 5.1** Three different screw configurations. **a** Screw configuration A: right handed screw with two kneading disc blocks, **b** screw configuration B: right handed screw with three kneading disc blocks and **c** screw configuration C: right handed screw followed one left handed screw with two kneading disc blocks

*Screw configurations*: Three different screw configurations were evaluated with the screw speed 150 rpm, the shear intensity of mixing was changed by changing the screw configuration. Figures 5.1 and 5.2 shows the screw configurations used in the study. The co-rotating screw configurations used in the study differed in the number of kneading disc blocks in the screw. The low shear intensity screw had one kneading block. The medium shear intensity configuration had two kneading blocks, and the high shear intensity configuration incorporated two kneading blocks and one reverse screw element. Tables 5.1 and 5.2 give basic configurations of screw element and distance in this study. Within a kneading block section, the inclusion of a reverse or neutral kneading element in the kneading block section causes an element of polymer to spend more time in the kneading block and increases the shear intensity of the screw configuration. Residence time, mechanical properties and morphology for each screw configurations were investigated with three different screw configuration (A, B, and C). PP/WF composites were prepared with PP 65 phr, PP-g-MA 5 phr, Wood–fiber 30 phr, and SEBS-g-MA 5 phr, respectively.

*Screw speeds*: Residence time is very important for extrusion process. Residence time directly depends on the screw speed. In order to determine the effect of screw speed on the mechanical and foaming properties, the WPC composite and screw configuration were hold constant and the screw speed varied, as shown in Table 5.3.

*Silica content*: the effect of different silica content on the mechanical and foaming properties was studied holding the screw configuration (C) and screw speed (150 rpm) constant as shown in Table 5.4.

*Compatibilizers*: the effect of compatibilizer on the mechanical and foaming properties using of PP/Wood–fiber (70/30) composite was studied holding the screw configuration (C) and screw speed (150 rpm) constant as shown in Table 5.5.

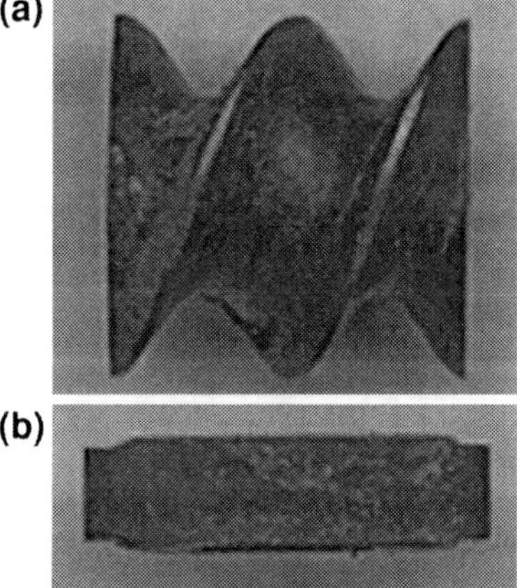


**Fig. 5.2** Pictorial views of **a** screw elements ([a]19 mm) and **b** kneading disc ([b]4.75 mm) of extruders

**Table 5.1** Basic configurations of screw element

| Designation | Screw configuration A | Screw configuration B | Screw configuration C |
|---|---|---|---|
| Right handed screw elements[a] | 34 | 32 | 31 |
| Left handed screw elements[a] | – | – | 1 |
| Right handed kneading discs[b] | 24 | 26 | 26 |
| Neutral kneading discs[b] | – | 6 | 6 |

[a] 19 mm and [b] 4.75 mm

**Table 5.2** Distance of screw element from hopper

| Screw configuration | A | B | C |
|---|---|---|---|
| Distance from hopper to first kneading block (mm) | 237.5 | 237.5 | 237.5 |
| Distance from hopper to second kneading block (mm) | – | 441.5 | 441.5 |
| Distance from hopper to third kneading block (mm) | 574.5 | 574.5 | 574.5 |
| Distance from hopper to second reverse element (mm) | – | – | 607 |
| Distance from hopper to die (mm) | 760 | 760 | 760 |

**Table 5.3** Formulation of wood–fiber/PP composite (effect of screw speeds)

| No. | Optimized parameters | Screw speed (rpm) |
|---|---|---|
| 1 | | 50 |
| 2 | Screw configuration C | 100 |
| 3 | Recipe: PP 65, wood-fiber 30, PP-g-MA 5, SEBS-g-MA 5 | 150 |
| 4 | | 200 |

**Table 5.4** Formulations of wood–fiber/PP composite (effect of silica content)

| No. | Optimized parameters | Various silica content (phr) |
|---|---|---|
| 1 | | 0.5 |
| 2 | Screw configuration C | 1 |
| 3 | Screw speed 150 rpm | 2 |
| 4 | Recipe: PP 65, wood-fiber 30, PP-g-MA 5, SEBS-g-MA 5 | 3 |
| 5 | | 5 |

**Table 5.5** Formulations of wood–fiber/PP composite (effect of compatibilizers)

| Sample | Optimized parameters | Various compatibilizer (10 phr) |
|---|---|---|
| A | | – |
| B | Screw configuration C | PP-g-MA |
| C | Screw Speed 150 rpm | SEBS |
| D | PP 70, wood-fiber 30 phr | SEBS-g-MA |

### *5.2.2 Preparation and Analysis of Wood–fiber/PP Composite Foams*

The PP/Wood–fiber composite after extrusion was pelletized, the plate samples of the composites with 2.0 mm thickness were compression-molding at 180°C for 6 min.

Microcellular foaming experiments were performed in a batch process. A schematic of the batch-foaming process is shown in Fig. 5.3. Plate samples that were 2.0 mm thick, 60.0 mm long, and 4.0 mm wide were enclosed high-pressure vessel. The vessel was flushed with low-pressure $CO_2$ for about 3 min and pressurized to the saturated vapor pressure $CO_2$ at room temperature and preheated to desired temperature. Afterward, the pressure was increased to the desired pressure by a syringe pump and maintained at this pressure for 1 h to ensure equilibrium absorption of $CO_2$ by the samples. After saturation, the pressure was quenched atmospheric pressure within 3 s and the samples were taken out. Then foam structure was allowed to full growth during rapid depressurization. The processing condition is listed in Table 5.2.

The average cell size and cell density were analyzed by utilizing the Image J software. The cell sizes, cell densities and relative densities were characterized. The cell diameter ($D$) is the average of all the cells on the SEM photo, usually more than 100 cells were measured.

$$D = \frac{\sum d_i n_i}{\sum n_i} \tag{5.1}$$

where $n_i$ was the number of cells with a perimeter-equivalent diameter of $d_i$.

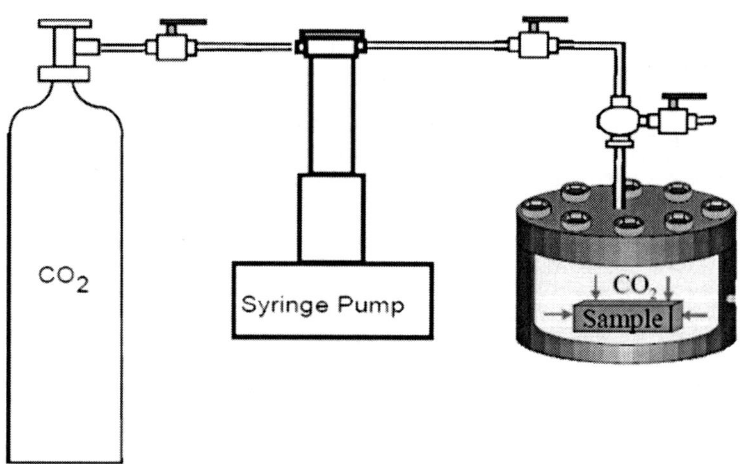


**Fig. 5.3** The schematics of batch-foaming process

The density of foam and unfoamed samples was determined from the sample weight in air and water, respectively, according to ASTM D 792 method A. Then the density of the foamed sample is divided by the density of the unfoamed sample to obtain the relative density ($\rho_r$). The volume fraction occupied by the microvoids ($V_f$) was calculated as

$$V_f = 1 - \frac{\rho_f}{\rho_m} \tag{5.2}$$

where $\rho_m$ and $\rho_f$ are the density of the unfoamed polymer and foamed polymer, respectively.

The cell density ($N_0$) based on the unfoamed sample was calculated as

$$N_f = \frac{V_f}{\frac{\pi}{6}D^3} \tag{5.3}$$

$$N_0 = \frac{N_f}{1 - V_f} \tag{5.4}$$

where $V_f$ is the volume fraction occupied by the microvoids, $N_f$ is the cell density based on the foamed sample.

## 5.3 Effect of Screw Configurations

In order to investigate the WF particles dispersion for each screw configuration, residence time, foaming properties, mechanical properties and morphology for each screw configurations were investigated with three different screw configuration (A, B, and C). The results of the residence time for each screw configuration are shown in Fig. 5.4. We found that the residence time increased with respect to the screw configurations in the order A < B < C which implies that the number of kneading disc block and reverse screw elements is a very important factor. The mechanical properties of the material extruded with different screw configurations are shown in Fig. 5.5. The Tensile strength and the elongation at break increase with the change of screw configurations in the order A < B < C. The mechanical properties of the PP/WF composite using configurations C is better than those produced using screw configurations A and B due to the longer residence time and good dispersion. The results of this study also showed a clear role of both kneading discs and left handed screw elements in the machine. The kneading discs are normally used to force the melting of the polymeric phases to occur early in its history inside the extruder to carry out a vigorous rapid dispersive mixing of ingredients. The position of the kneading discs allows determining the exact location of the melting and intensive mixing. Introduction of the additional kneading blocks other than the normal kneading discs for melting and initial dispersion causes the dispersive mixing to proceed rapidly resulting in better

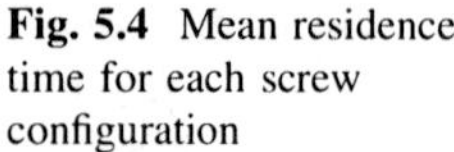

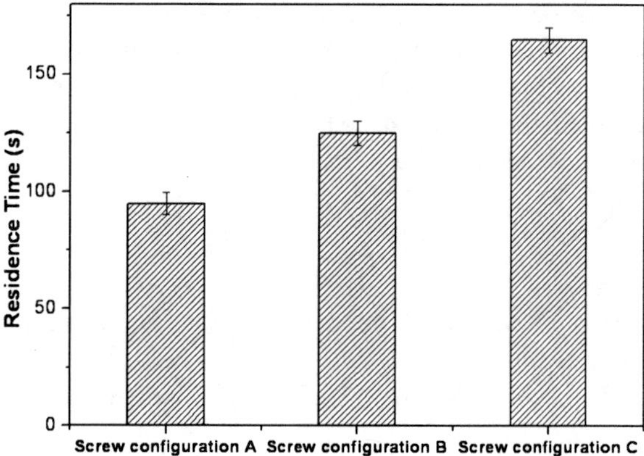


**Fig. 5.4** Mean residence time for each screw configuration

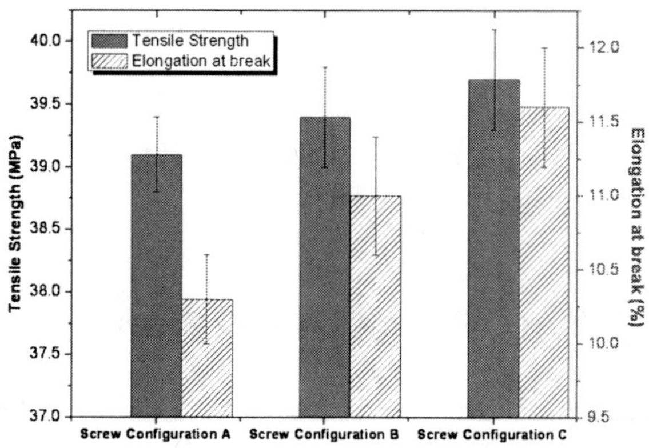


**Fig. 5.5** The mechanical properties with different screw configurations

mechanical properties. This can be easily seen when Screw configuration B which has additional kneading blocks (with other configurations almost same) has better properties than A. The left handed screw elements functions to control the pressure field and can cause the development of fully filled regions.

The results obtained by compiling the comparison of different screw geometries show different morphologies of the blends. The results of mechanical properties are well supported by the SEM morphology as shown in Fig. 5.6a–c. Since the residence time of the blends in screw configuration A and B is very short, meaning there is less time and lower shear rate for compatibilization to occur thereby resulting in lower mechanical properties. This is attributed to the design of the screw design configuration.

In the case of immiscible polymer blends, its morphology determines the physical properties. Morphological changes are very dependent on local flow conditions (shear rate, temperature and residence time) which will modify the material properties (viscosity, interfacial tension coefficient) and that the final distribution obtained at the die exit is the result of a long and complex process [30].

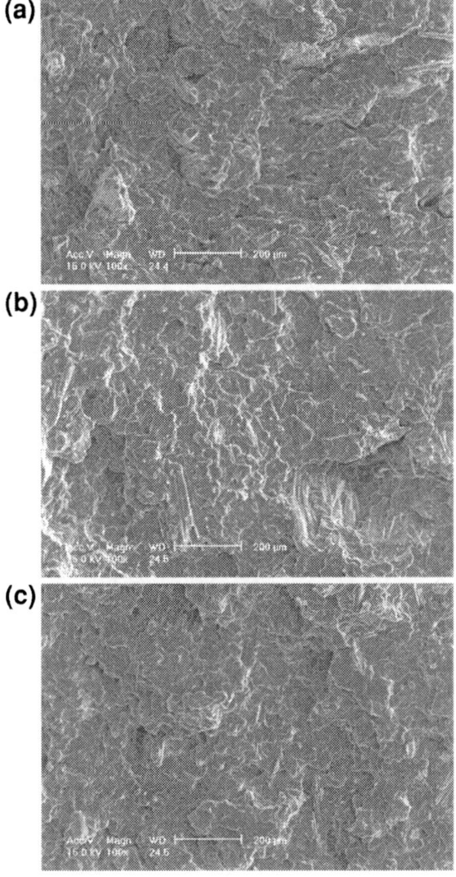


**Fig. 5.6** SEM photographs of samples prepared at different screw configurations. **a** Screw configuration A, **b** screw configuration B, **c** screw configuration C

From the SEM micrographs it is evident that the morphologies of the fractured surfaces follows two distinct patterns with configurations A and B showing in-homogeneity while that of C showing better homogeneity. For configuration A, it can be seen that a large number of WF particles have pulled out, in case of configuration B is less but almost absent in C. This means that the interfacial adhesion between PP and WF particles is so poor in blends extruded by screw A. Moreover, the samples extruded by C shows a smoother surface indicating good adhesion. This is because the reverse pumping element in screw C ensures the mix more homogeneous and reaction more sufficient between MA groups in compatibilizers and –OH groups in WF [9]. The reaction taking place within the composites during processing is shown in Fig. 5.7. Thus, we can conclude that the dispersion of the particles increases in the order of screw configuration, A < B < C.

Batch foaming experiments were performed to study the effect of screw configuration on the foamed cell morphology. Figure 5.8 shows the scanning electron micrographs for foamed PP/WF composites that were saturated with $CO_2$ at 155°C

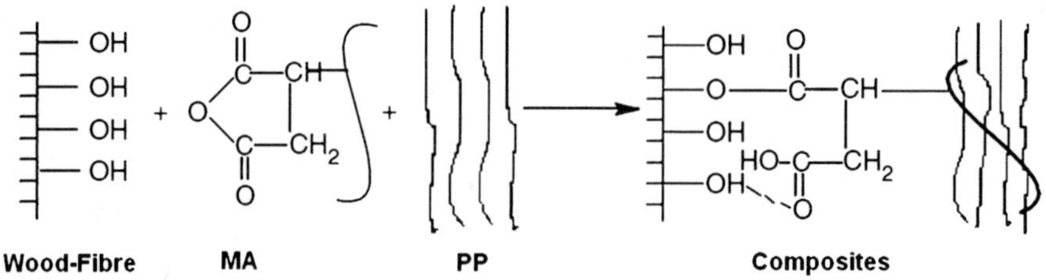

Wood-Fibre MA PP Composites

Fig. 5.7 Chemical reactions between functional groups [9]

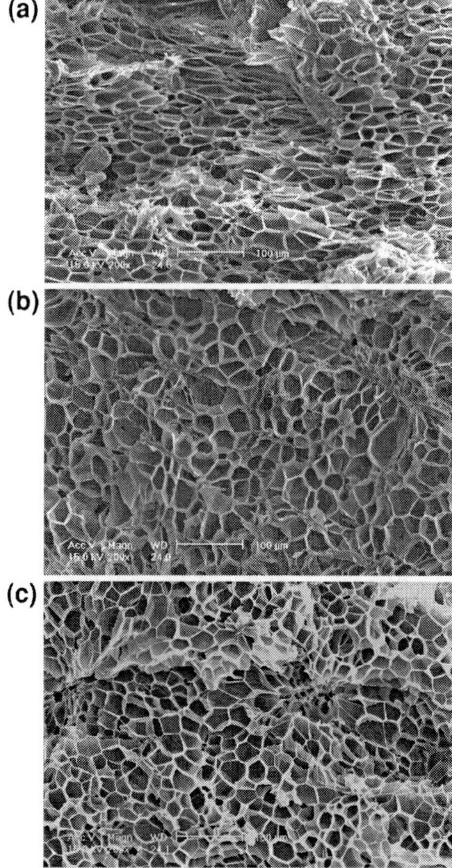


**Fig. 5.8** Scanning electron micrographs of foamed PP/WF composites were prepared at different screw configurations. **a** Screw configuration A, **b** screw configuration B, **c** screw configuration C

and 16 MPa for 1 h and then the pressure was quenched atmospheric pressure within 3 s. The composite foams were analyzed in terms of cell size distribution. From Fig. 5.8, it is clear that screw configuration have definite effect on cell size and cell size distribution. Figure 5.9 shows the effect of screw configuration on the

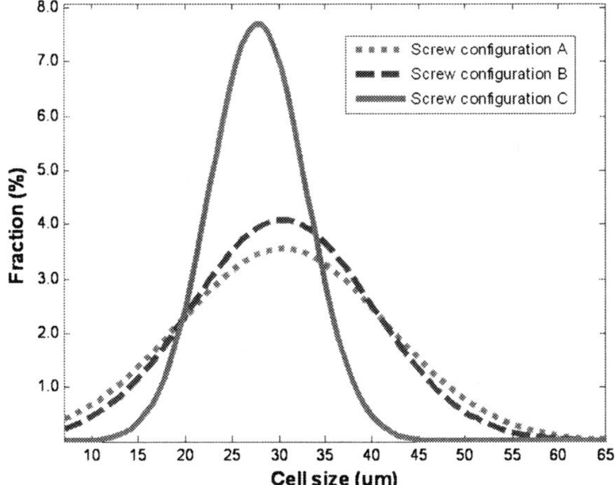


**Fig. 5.9** Effect of screw configuration on cell size distribution of foamed wood–fiber/PP composites

cell size distribution. Under similar foaming conditions, the cell size of PP/WF composite produced at screw configuration C is smaller than that of composite foams produced at screw configuration A and B, and composite foams produced at screw configuration shows narrower cell size distribution, namely, the cell morphology of composite foamed at screw configuration improved. This improvement can be attributed to the dispersed wood fiber in the polymer matrix. These particles are believed to act like nucleating agent, thus promoting heterogeneous nucleation. At the same foaming conditions, the amount of gas available for foaming is constant. Increasing the number cell through heterogeneous nucleation leads to smaller sizes.

## 5.4 Effect of Screw Speed

Figure 5.10 shows the mechanical properties of the sample blends extruded in different screw speeds (Table 5.3) using the optimum screw configuration C. The PP/WF composites extruded at 150 rpm showed the maximum mechanical properties viz. tensile strength and elongation at break compared to 50, 100, 150 and 200 rpm. This is due to higher shear rate obtained at 150 rpm screw speed than 50 and 100 rpm, and higher residence time than the screw speed of 200 rpm, as shown in Fig. 5.11. Figure 5.12 shows the morphology of the samples prepared at different screw speed in screw configuration C. The morphology at screw rpm of 150 was the best than other screw speed. The technical importance of reaction to make the PP/WF composite needs high shear stress at PP melt state and suitable residence time. Based on experimental results, the best screw speed is 150 rpm for screw configuration C because this is the best condition for reaction to occur. These results indicate that the properties strongly depend on the reaction between maleic anhydride groups of compatibilizers and OH groups on WF particles during processing.

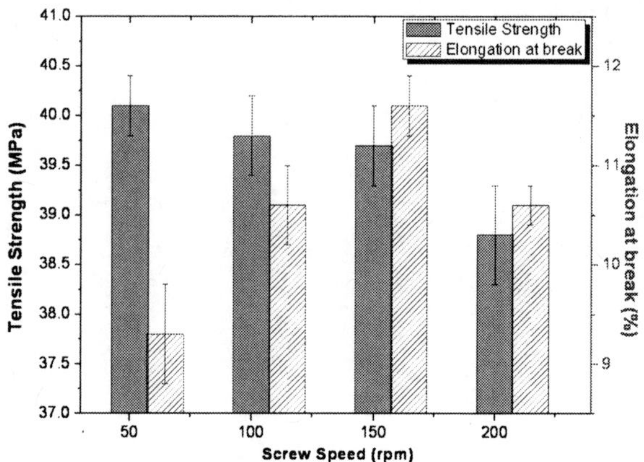


**Fig. 5.10** The tensile strength and elongation at break of the PP/WF composite extruded with different screw speeds using screw configuration C

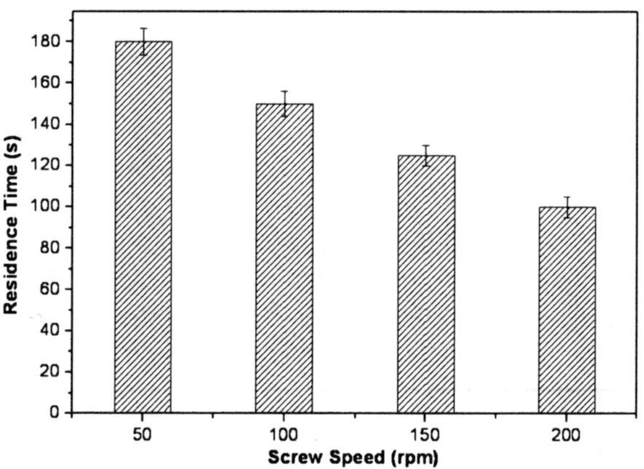


**Fig. 5.11** Mean residence time at different screw speeds in screw configuration C

Batch foaming experiments were performed to study the effect of screw speed on the foamed cell morphology. Figure 5.13 shows the scanning electron micrographs for foamed PP/WF composites that were saturated with $CO_2$ at 155°C and 16 MPa for 1 h and then the pressure was quenched atmospheric pressure within 3 s. The composite foams were analyzed in terms of cell size distribution. From Fig. 5.13, it is can be seen that screw speed has not too much effect on cell morphology. Figure 5.14 shows the effect of screw configuration on the cell size distribution. Under similar foaming conditions, the cell size of PP/WF composite produced at screw speed of 150 is a little smaller than that of composite foams produced at other screw speeds, and composite foams produced at screw speed of 150 and 200 show narrower cell size distribution, namely, the cell morphology of composite foamed at screw speed of 150 and 200 improved. This improvement can be attributed to the dispersed wood fiber in the polymer matrix. As aforementioned, the WF particles are believed to act like nucleating agent, thus promoting

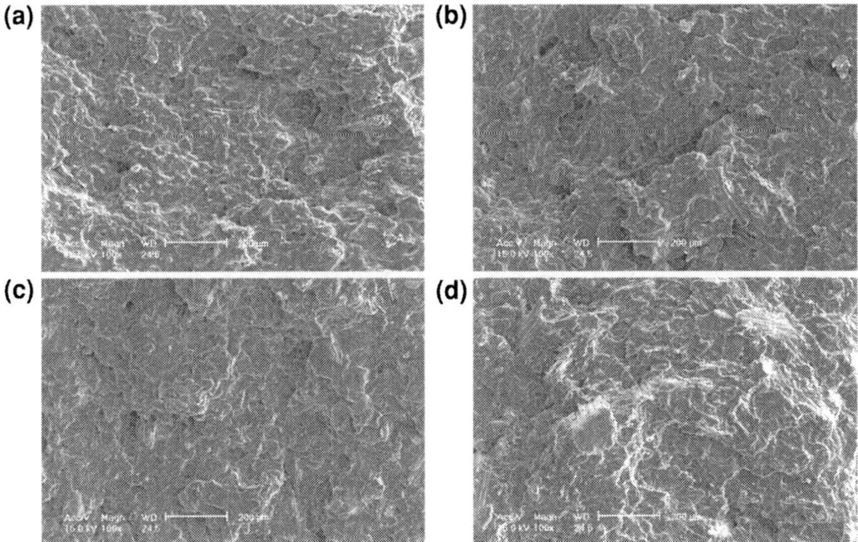

**Fig. 5.12** SEM photographs of samples prepared at different screw speeds in screw configuration D. **a** 50 rpm, **b** 100 rpm, **c** 150 rpm, **d** 200 rpm

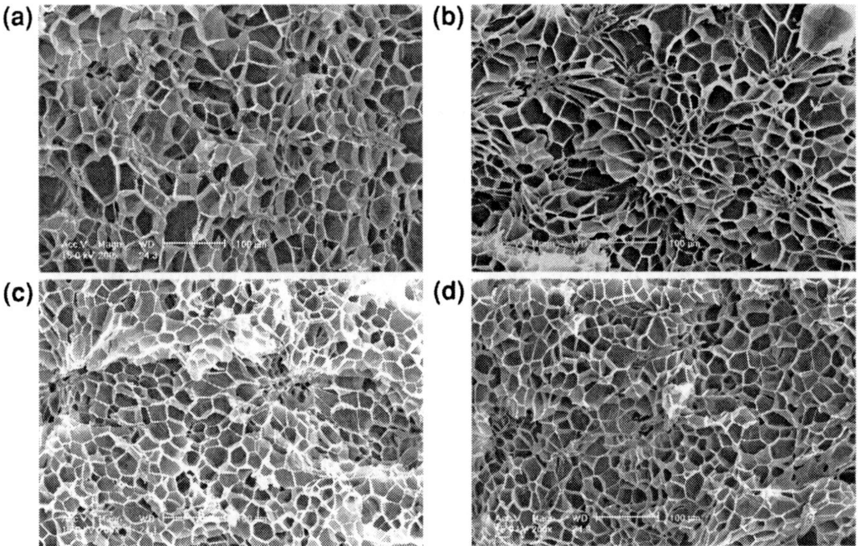

**Fig. 5.13** Scanning electron micrographs of foamed PP/WF composites were prepared at different screw configurations. **a** 50 rpm, **b** 100 rpm, **c** 150 rpm, **d** 200 rpm

heterogeneous nucleation. At the same foaming conditions, the amount of gas available for foaming is constant. Increasing the number cell through heterogeneous nucleation leads to smaller sizes.

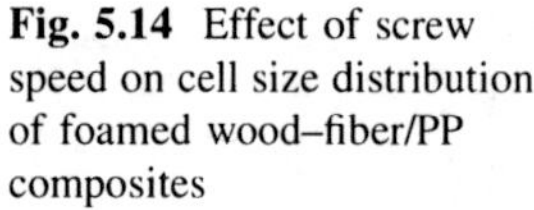

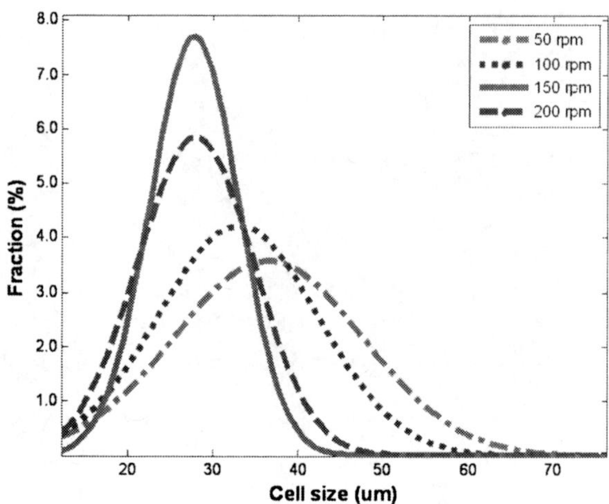


**Fig. 5.14** Effect of screw speed on cell size distribution of foamed wood–fiber/PP composites

## 5.5 Effect of Silica Content

In order to investigate the silica content on the mechanical and foaming properties of PP/WF composite, mechanical properties, morphology and foaming properties were investigated with different content of silica (0.5, 1, 2, 3, 5 phr).

The mechanical properties, tensile strength, and elongation at break with different content of fumed silica are shown in Fig. 5.15. Silica is usually used as an enhancing agent in thermoplastic polymers to increase the mechanical properties, such as tensile strength and toughness. It can be seen that both the tensile strength and elongation at break increased for silica content up to 2 phr, for high concentrations, the reduction was observed in the samples, this trend should be associated with the extended aggregation of silica nonoparticles, which, as observed in micrographs, increased with increasing silica content. This finding was in agreement with a similar deterioration in mechanical properties reported for PP/$SiO_2$ nanocomposites [31]. From Fig. 5.16, impact strength of composite with

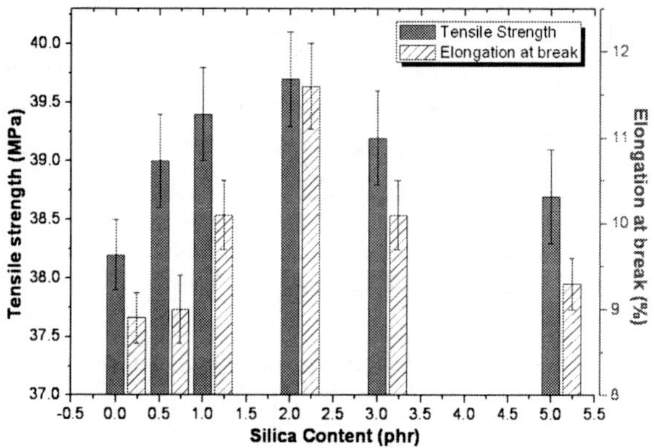


**Fig. 5.15** The mechanical properties of PP/WF composites with different content of silica

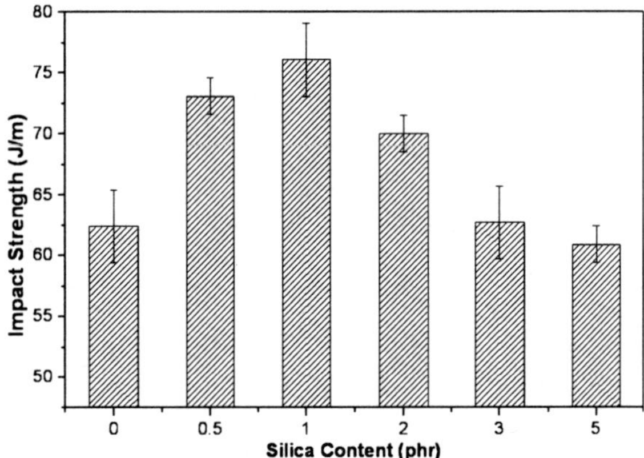


**Fig. 5.16** The impact strength of PP/WF composites with different content of silica

addition of small content silica was found to be higher than of PP/WF composite without silica, like other mechanical properties, apparently the presence of large silica agglomerates reduced impact strength.

It is well known that filler dispersion and adhesion with the polymer matrix are of great importance in improving the mechanical behavior of composites. Fine control of the interface morphology of polymer nanocomposites are one of the parameters most critical to imparting the desired mechanical properties of such materials. It can be seen that the silica agglomerates shown in Fig. 5.17 are well distinguished. From these micrographs it can be seen that fumed silica at 3 and 5 phr created big agglomerates. For this reason, particle dispersion into the matrix was not homogeneous; this is also the evidence of the reason of reduction in mechanical properties.

Batch foaming experiments were performed to study the effect of silica on the foamed cell morphology. Figure 5.18 shows the scanning electron micrographs for foamed PP/WF composites that were saturated with $CO_2$ at 150°C and 20 MPa for 1 h and then the pressure was quenched atmospheric pressure within 3 s. The resulting foams were radically different. Compared with foams without silica, the PP/WF/silica composite foams have smaller cell size and a more homogeneous cell distribution. The difference in the resulting cellular structures of the samples seemed to be due to the fact that the silica not only played a role as heterogeneous nucleating agents, thus providing more sites for cell nucleation, but also prevented cell coalescence during cell growth, which in turn led to an increase in cell density [32]. In the heterogeneous nucleation scheme, the activation energy barrier to nucleation is sharply reduced in the presence of filler particles [33, 34]. However, when the concentration of silica increases up to 3 phr, coalescence of cells appears, and it becomes more obvious for the 5 phr silica sample. A possible reason, as aforementioned that fumed silica at 3 and 5 phr created big agglomerates, which caused the cell coalescence, and a number of big bubble occurred. More $CO_2$ gas diffused out of the polymer matrix because of the increased $CO_2$ diffusivity. As shown in Fig. 5.19, the cell size decreases and cell density increases with a small content of silica addition and then levels off at high silica

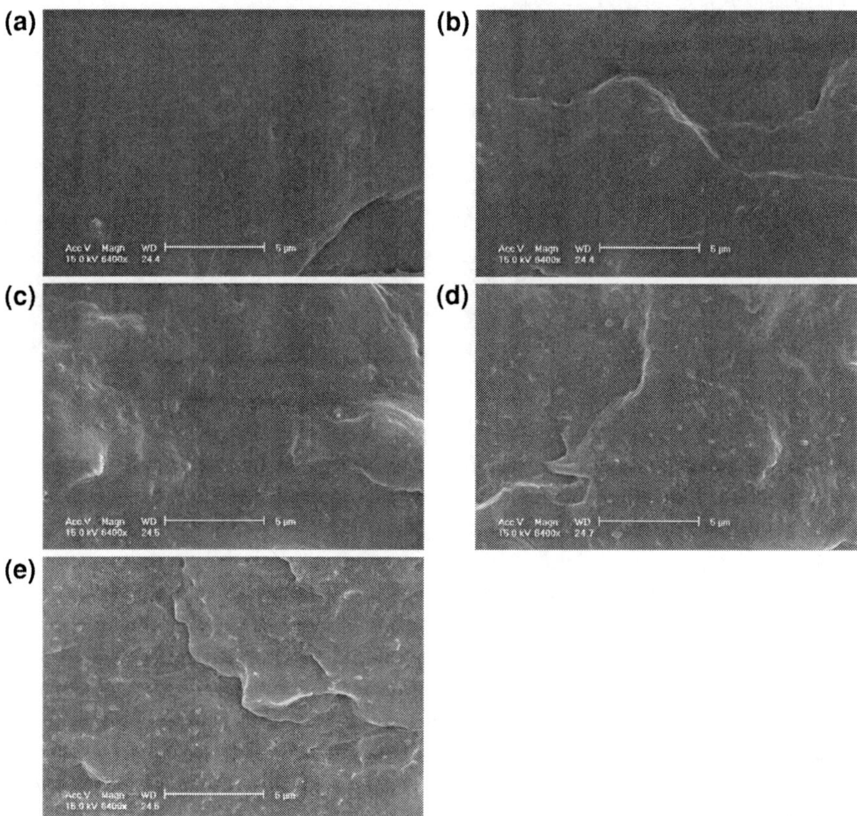


**Fig. 5.17** SEM micrographs of fractured surface of PP/WF/Silica nanocomposites with different content of silica. **a** 0.5 phr, **b** 1 phr, **c** 2 phr, **d** 3 phr, **e** 5 phr

concentration. In Fig. 5.20, the relative density decreases with increase of silica up to 2 phr content and then increases. It is believed that the relative density is the completion between the cell nucleation and the cell growth. The relative density is also increased when the silica content more than 2 phr due to increase of mass by the addition of silica.

## 5.6 Effect of Various Compatibilizers

### *5.6.1 Effect of Compatibilizer on Crystallinity*

The results of the effect of compatibilizers on crystallinity of PP/WF composites are summarized in Table 5.6. Random PP is a semi-crystalline polymer, and its mechanical properties are greatly affected by its overall crystallinity, the foaming

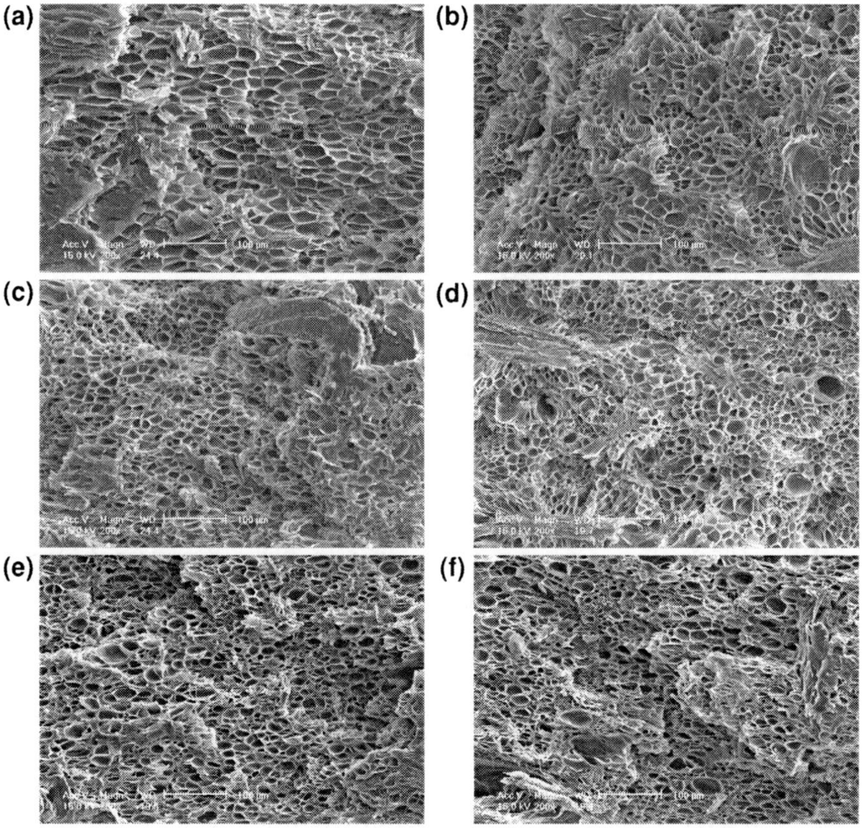

**Fig. 5.18** Scanning electron micrographs of foamed PP/WF/Silica nanocomposites at a saturation temperature of 150°C and a pressure of 20 MPa: **a** 0 phr, **b** 0.5 phr, **c** 1 phr, **d** 2 phr, **e** 3 phr, **f** 5 phr

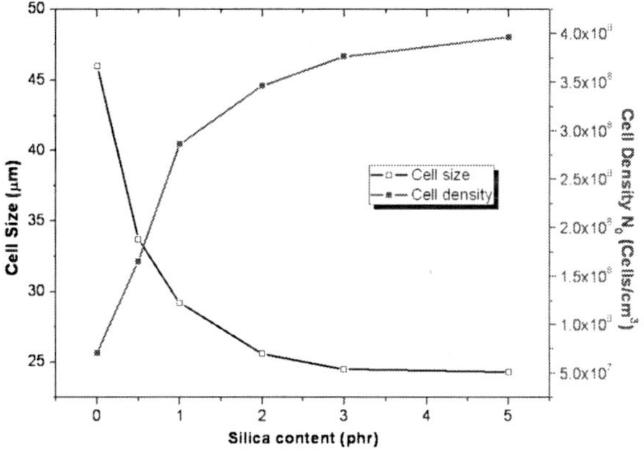


**Fig. 5.19** Effect of silica content on the cell size and cell density of foamed PP/WF/Silica nanocomposites at a saturation temperature of 150°C and a pressure of 20 MPa

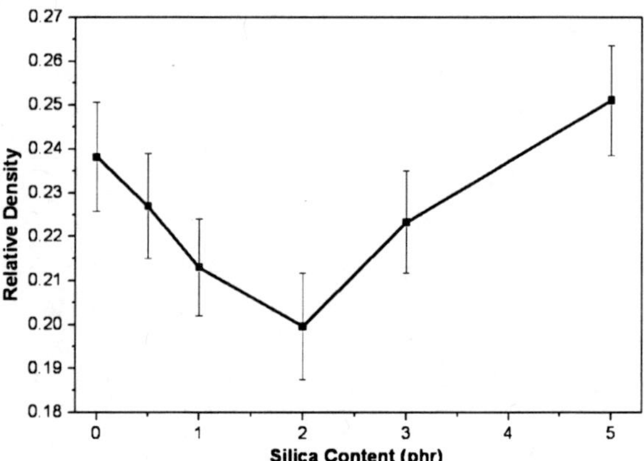


**Fig. 5.20** Effect of silica content on the relative density of foamed PP/WF/Silica nanocomposites at a saturation temperature of 150°C and a pressure of 20 MPa

**Table 5.6** Crystallization temperature, crystallinity of PP/WF composites

| Sample | $T_c$ (°C) | Crystallinity of composites (%) |
|---|---|---|
| PP | 125.33 | 41.15 |
| PP/wood (70/30) | 112.5 | 22.04 |
| PP/wood/PP-g-MA (70/30/10) | 117.60 | 22.67 |
| PP/wood/SEBS-g-MA (70/30/10) | 110.54 | 20.85 |
| PP/wood/SEBS (70/30/10) | 113.4 | 19.39 |

behavior of a semi-crystalline polymer largely depends on the crystallization temperature [35]. Therefore, it is important to investigate the effects of various compatibilizers on the crystallization behavior of blends. The crystallization temperature ($T_c$) and crystallinity of PP/WF composites decreased when the addition of wood fiber and compatibilizers, except for the PP-g-MA as compatibilizer. This effect could be because of the lowered mobility of polymer chains in the PP/WF composite. And the compatibilized compounds, compatibilization with SEBS-g-MA and SEBS caused a decrease in crystallization temperature, because this kind of compatibilizer acting as a retarding agent in the composites. Whereas the blend containing PP-g-MA as compatibilizers exhibited increased in crystallization temperature, this may have been caused by the PP-g-MA acting as nucleating agents in the blend.

### 5.6.2 *Effect of Compatibilizers on Rheological Properties of PP/WF Composites*

Plots of shear viscosity versus shear rate for the blends at 180°C are exhibited in Fig. 5.21. All the materials behave in a non-Newtonian manner. It is also obvious that all the WPC composites under study exhibited shear-shinning behavior. From

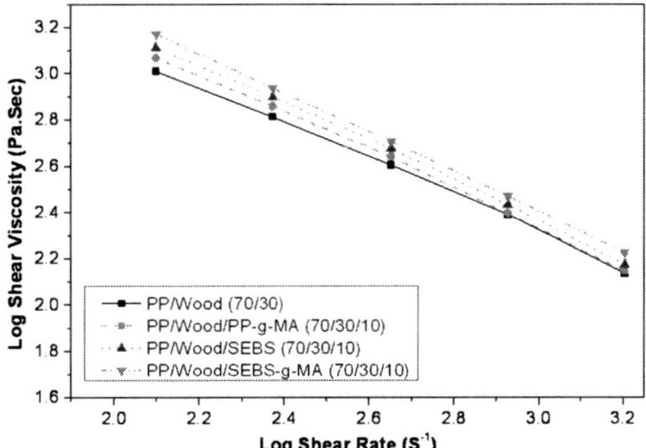


**Fig. 5.21** Shear viscosities of PP/WF composites at 180°C

Fig. 5.21, it can be seen that the shear viscosity of PP/WF composites increases in the presence of compatibilizer due to the occurrence of reaction between wood and polar compatibilizer or the increase of compatibility between wood fiber and nonpolar compatibilizer. Moreover, the shear viscosity of WPC composites is almost the largest in presence of SEBS-g-MA, followed by SEBS, PP-g-MA and without compatibilizer.

### *5.6.3 Effect of Compatibilizer on the Mechanical Properties and Morphology of PP/WF Composites*

Figure 5.22 shows the tensile and elongation of PP/WF composite (bar 1) compared with composites with different compatibilizer. The tensile strength is highest for the PP/WF composite with PP-g-MA as a compatibilizer, the composites with SEBS-g-MA also have increased the tensile strength. The result is expected because of the PP-g-MA and SEBS-g-MA can improve the interracial bonding between WF and PP resulting in improved tensile strength, whereas the SEBS had a negative effect on the tensile strength. The elongation at break is greater for all composites with compatibilizer is added, the SEBS-g-MA composite systems have the highest elongation at break, this result is also proved by the highest shear viscosity of composite with the addition of SEBS-g-MA. Figure 5.23 shows that the stiffness of PP/WF composites. It can be observed that all compatibilizers decrease the stiffness of the composites except for the PP-g-MA, this is due to the low modulus of the elastomer. The PP-g-MA composite shows the highest tensile modulus, and decrease in the order of PP-g-MA > without compatibilizer > SEBS-g-MA > SEBS. Figure 5.24 shows the impact strength of each composition. Improvements in impact strength were seen in all composition system. The addition of SEBS-g-MA resulted in greater improvements in impact

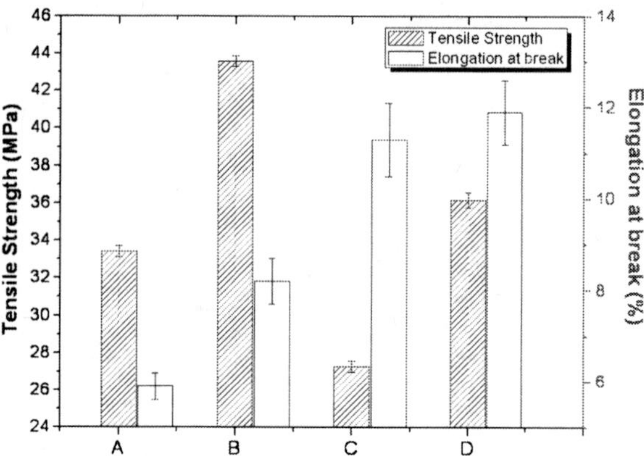


**Fig. 5.22** Effect of various compatibilizer on the tensile and elongation at break. **a** No compatibilizer, **b** PP-g-MA, **c** SEBS, **d** SEBS-g-MA

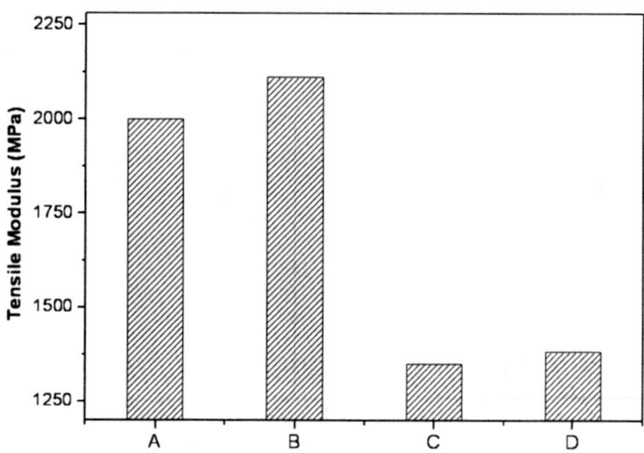


**Fig. 5.23** Effect of compatibilizers on the modulus of the PP/WF composites. **a** No compatibilizer, **b** PP-g-MA, **c** SEBS, **d** SEBS-g-MA

performance. The PP-g-MA compatibilizer had little or no effect on the notched impact strength.

Examination of the fracture surfaces of the composites by SEM gave information about how compatibilizers affect the morphology of the composite. Figure 5.25a shows the microstructure of the composite without compatibilizer showing wood particle embedded in the polymer matrix. There are big voids around the particle indicating poor interaction between the wood surface and the PP matrix. Figure 5.25b, d shows the morphology of the composite with PP-g-MA and SEBS-g-MA. There is good adhesion between the wood fiber and PP matrix in both micrographs. No voids around the wood particle surfaces are present. Almost all mechanical properties were improved compared to other composites, as shown in Figs. 5.22, 5.23, 5.24. Figure 5.25c show the micrographs of PP/WF composites with SEBS. There is poor adhesion between the PP matrix and WF particle surface, as there are voids around the wood particle, this is also the evidence of the poor mechanical properties of the PP/WF composites with SEBS as compatibilizer.

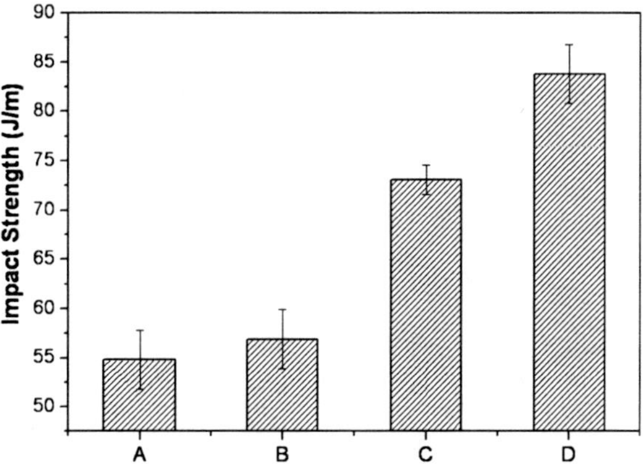


**Fig. 5.24** Effect of compatibilizers on the impact strength of the PP/WF composites. **a** No compatibilizer, **b** PP-g-MA, **c** SEBS, **d** SEBS-g-MA

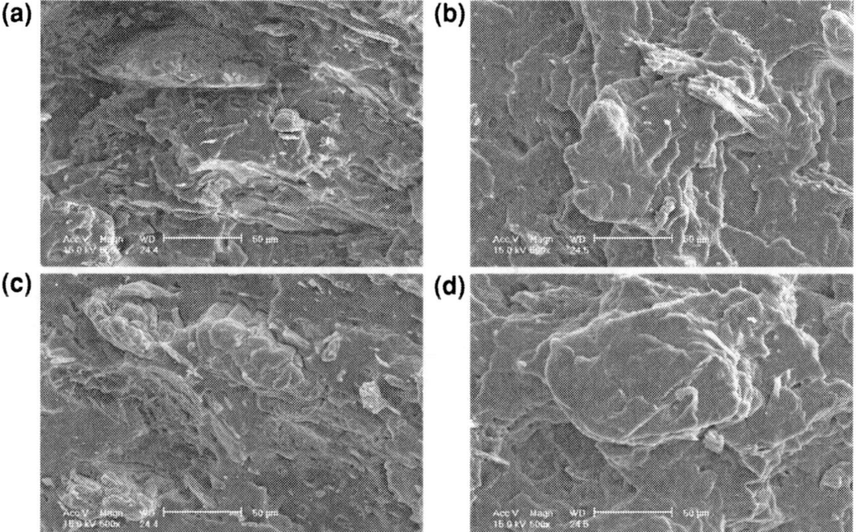


**Fig. 5.25** SEM micrograph of fractured PP/WF composites with various compatibilizer. **a** No compatibilizer, **b** PP-g-MA, **c** SEBS, **d** SEBS-g-MA

### *5.6.4 Effect of Compatibilizer on the Foaming Properties of PP/WF Composites*

Batch foaming experiments were performed to study the effect of compatibilizer on the foamed cell morphology. Figure 5.26 shows the scanning electron micrographs for foamed PP/WF composites that were saturated with $CO_2$ at 150°C and 16 MPa for 1 h and then the pressure was quenched atmospheric pressure within 3 s. The cell structures were radically different. It can be observed that without

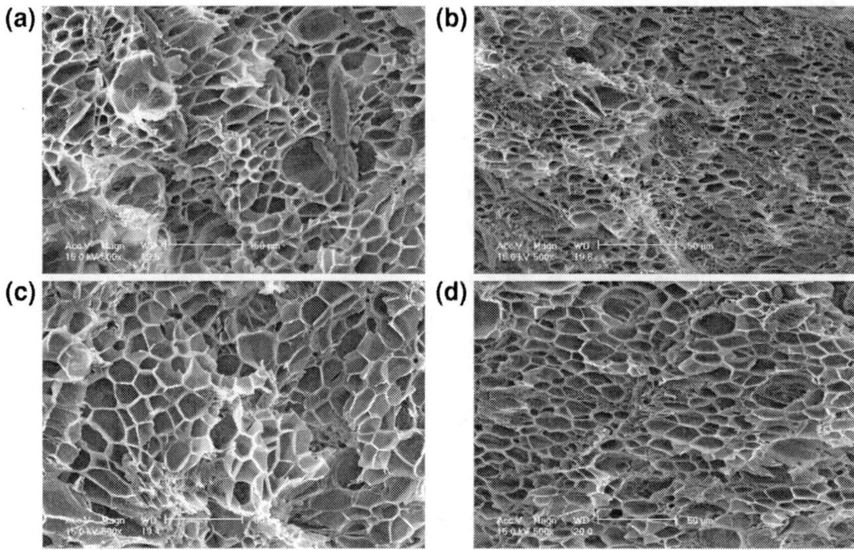

**Fig. 5.26** Scanning electron micrographs of foamed PP/WF composites with different compatibilizer at a saturation temperature of 150°C and a pressure of 16 MPa: **a** No compatibilizer, **b** PP-g-MA, **c** SEBS, **d** SEBS-g-MA

compatibilizer, PP/WF composite produced foams with some large bubbles and non-uniform cell structure, this was expected because the poor adhesion between WF particle and PP matrix as shown in Fig. 5.25a. PP/WF composite with PP-g-MA as compatibilizer produced foams with significantly smaller cell size and shows bimodal cellular structure. Figures 5.27 and 5.28 shows the cell size, cell density and relative density of PP/WF composite with different compatibilizer, the PP/WF composite with PP-g-MA as a compatibilizer shows the highest relative density and cell density and smallest cell size, but the cell size is biggest, cell density and relative density is lowest for the PP/WF composite with SEBS as a compatibilizer. The difference in the resulting cellular structures and relative density of samples seems to be due to the different crystallization of the samples. One explanation is that the samples with a higher crystallinity received less gas during the saturation process because they had less amorphous regions. As shown in Table 5.6, the crystallinity of composite with different compatibilizer are consistent with the foaming result except for the PP/WF composite without compatibilizer, namely, lower crystallinity cause to bigger cell size and lower relative density, however, PP/WF composite without compatibilizer with second highest crystallinity but shows bigger cell size, this is because of, as aforementioned, the poor surface adhesion between WF particles and PP matrix caused partially large bubbles, so the average cell size increased. Moreover, the stiffness of the polymer matrix could also have played an important role in foaming [36]. With addition of rubber phase in the PP/WF composites also reduced the stiffness of the materials, which in turn increases the average cell size in the microstructure.

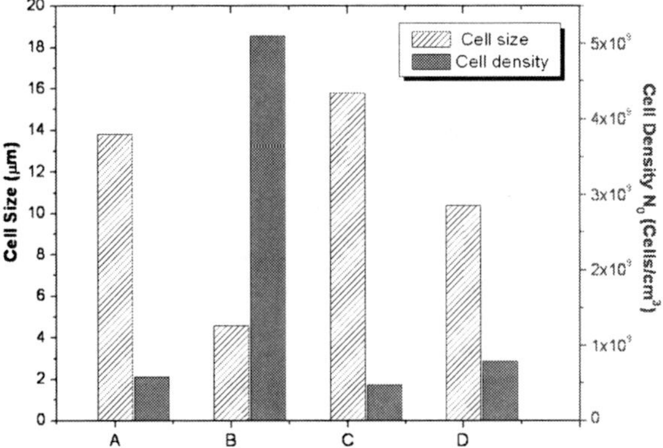


**Fig. 5.27** Effect of different compatibilizer on the cell size and cell density of foamed wood–fiber/PP composites at a saturation temperature of 150°C and a pressure of 16 MPa

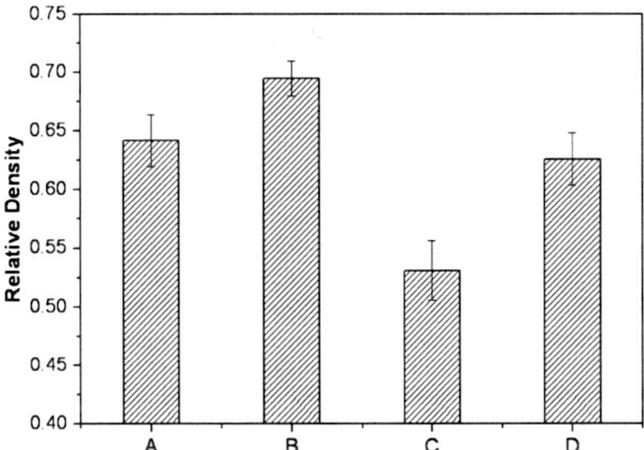


**Fig. 5.28** Effect of various compatibilizer on the relative density of foamed Wood–fiber/PP composites at a saturation temperature of 150°C and a pressure of 16 MPa

For example, as shown in Fig. 5.23, PP/WF composite with SEBS as compatibilizer has lowest stiffness which corresponding to the biggest cell size and relative density are shown in Figs. 5.27 and 5.28, respectively.

## References


1. Bledzki, A.K., Gassan, J.: Composites reinforced with cellulose based fibres. Prog. Polym. Sci. **24**(2), 221–274 (1999)
2. Bledzki, A.K., Faruk, O., Huque, M.: Physico-mechanical studies of wood fiber reinforced composites. Polym. Plast. Technol. Eng. **41**(3), 435–451 (2002)
3. Bledzki, A.K., Letman, M., Viksne, A., Rence, L.: A comparison of compounding processes and wood type for wood fibre-PP composites. Composites A**36**(6), 789–797 (2005)
4. Ichazo, M.N., Albano, C., Gonzalez, J., Perera, R., Candal, M.V.: Polypropylene/wood flour composites: treatments and properties. Compos. Struct. **54**(2–3), 207–214 (2001)
5. Bledzki, A.K., Reihmane, S., Gassan, J.: Thermoplastics reinforced with wood fillers: a literature review. Polym. Plast. Technol. Eng. **37**(4), 451–468 (1998)

6. Bledzki, A.K., Letman, M., Viksne, A., Rence, L.: A comparison of compounding processes and wood type for wood fibre—PP composites. Compos. Part A Appl. Sci. Manuf. **36**, 789–797 (2005)
7. Matuana L.M., Heiden, P.A.: Wood composites. In: Kroschwitz J.I. (ed.) Encyclopedia of polymer science and technology. Wiley, New York (2004)
8. Hristov, V.N., Krumova, M., Vasileva, St., Michler, G.H.: Modified polypropylene wood flour composites. II. Fracture, deformation, and mechanical properties. J. Appl. Polym. Sci. **92**(2), 1286–1292 (2004)
9. GUO, C.G., Wang, Q.W.: Compatibilizing effect of maleic anhydride grafted styrene-ethylene-butylene-styrene (MAH-g-SEBS) on the polypropylene and wood fiber composites. J. Reinf. Plast. Compos. **26**(17), 1743–1752 (2007)
10. Wu, J.S., Yu, D., Chan, C.M., Kim, J.G., Mai, Y.W.: Effect of fiber pretreatment condition on the interfacial strength and mechanical properties of wood fiber/PP composites. J. Appl. Polym. Sci. **76**, 1000–1010 (2000)
11. Nitz, H., Reichert, P., Romling, H., Mulhaupt, R.: Influence of compatibilizers on the surface hardness, water uptake and the mechanical properties of poly(propylene) wood flour composites prepared by reactive extrusion. Macromol. Mater. Eng. **276**, 51–58 (2000)
12. Danyadi, L., Janecska, T., Szabo, Z., Nagy, G., Moczo, J., Pukanszky, B.: Wood flour filled PP composites: compatibilization and adhesion. Compos. Sci. Technol. **67**(13), 2838–2846 (2007)
13. Kim, M.H., White, J.L.: Modelling flow in tangential counter-rotating twin screw extruders. Int. Polym. Proc. **3**, 201–207 (1990)
14. White, J.L.: Twin screw extrusion. Hanser, Munich, New York, 1990
15. Rauwendaal, C.: Mixing in polymer processing, Marcel Dekker, New York, 1991
16. Kye, H.S., White, J.L.: Continuous polymerization of caprolactam in a modular intermeshing corotating twin screw extruder integrated with continuous melt spinning of polyamide 6 fiber: influence of screw design and process conditions. J. Appl. Polym. Sci. **52**, 1249–1262 (1994)
17. Lachtermacher, M.G., Rudin, A.: Reactive processing of LLDPES in corotating intermeshing twin-screw extruder. II. Effect of peroxide treatment on processability. J. Appl. Polym. Sci. **58**, 2433–2449 (1995)
18. Kim, P.J., White, J.L.: Transesterification of ethylene vinyl acetate copolymer in a modular intermeshing corotating twin screw extruder with different screw configurations. J. Appl. Polym. Sci. **54**, 33–45 (1994)
19. Potente, H., Bastian, M., Gehring, A., Stephan, M., Pötschke, P.: Experimental investigation of the morphology development of polyblends in corotating twin-screw extruders. J. Appl. Polym. Sci. **76**, 708–721 (2000)
20. Schmidt, L.R., Lovgran, E.M., Meissner, P.G.: Int. Polym. Proc. **4**, 270 (1989)
21. Lyu, M.Y., White, J.L.: Models of flow and experimental studies on a modular list/Buss Kokneter. Int. Polym. Proc. **4**, 305–313 (1995)
22. Kim, J.K., Lee, S.H. and Hwang, S.H. 2001. A study on thermoplastic elastomer blend using waste rubber powder (I): screw configurations, morphologies and mechanical properties. 2001. Elastomer **36**(2), 86–93
23. Zhang, J., Park, C.B., Rizvi,G.M., Huang, H., Guo, Q.: Investigation on the uniformity of high-density polyethylene/wood fiber composites in a twin-screw extruder. J. Appl. Polym. Sci. **113**, 2081–2089 (2008)
24. Goel, S.K.; Beckman, E.J.: Generation of microcellular polymeric foams using supercritical carbon dioxide. II: Cell growth and skin formation. Polym. Eng. Sci. **34**, 1148–1156 (1994)
25. Han, X., Zeng, C., Lee, L.J., Koelling, K.W., Tomasko, D.L.: Extrusion of polystyrene nanocomposite foams with supercritical CO2. Polym. Eng. Sci. **43**, 1261–1275 (2003)
26. Shen, J., Zeng, C., Lee, L.J.: Synthesis of polystyrene–carbon nanofibers nanocomposite foams. Polymer **46**, 5218–5224 (2005)
27. Munoz, E., Val, M.D., Ruiz-Gonzalez, M.L., et al.: Gold/carbon nanocomposite foam. Chem. Phys. Lett. **420**, 86–89 (2006)

28. Lee, Y.H., Sain, M., Kuboki, T., Park, C.B.: Extrusion foaming of nano-clay-filled wood fiber composites for automotive applications. Int. J. Mater. Manuf. **1**(1), 641–647 (2009)
29. Lee, Y.H., Kuboki, T., Park, C.B., Sain, M.: Effect of nanoclay on extrusion foaming of WF/PP/Clay composites using N2. AIChE Annual Meeting, Philadelphia, PA, 16–21 Nov 2008
30. De Loor, A., Cassagnau, P., Michel, A., Vergnes, B.: Morphological changes of a polymer blend into a twin-screw extruder. Int. Poylm. Process. **IX**(3), 211–218 (1994)
31. Bikiaris, D.N., Papageorgiou, G.Z., Pavlidou, E., Vouroutzis, N., Palatzoglou, P., Karayannidis, G.P.: Preparation by melt xixing and characterization of isotactic polypropylene/SiO2 nanocomposites containing untreated and surface-treated nanoparticles. J. Appl. Polym. Sci. **100**, 2684–2696 (2006)
32. Lee, Y.H., Park, C.B., Wang, K.H., Lee, M.H.: HDPE-clay nanocomposite foams blown with supercritical CO2. J. Cell. Plast. **41**, 487–502 (2005)
33. Guo, G., Wang, K.H., Park, C.B., Kim, Y.S., Li, G.: Effects of nanoparticles on the density reduction and cell morphology of extruded metallocene polyethylene/wood fiber nanocomposites. J. Appl. Polym. Sci. **104**, 1058–1063 (2006)
34. Wee, D., Seong, D.G., Youn, J.R.: Processing of microcellular nanocomposite foams by using a supercritical fluid. Fibers Polym. **5**(2), 160–169 (2004)
35. Doroudiani, S., Park, C.B., Kortschot, M.T.: Effect of the crystallinity and morphology on the microcellular foam structure of semicrystalline polymers. Polym. Eng. Sci. **36**(21), 2645–2662 (1996)
36. Wong, S., Naguib, H.E., Park, C.B.: Effect of processing parameters on the cellular morphology and mechanical properties of thermoplastic polyolefin (TPO) microcellular foams. Adv. Polym. Technol. **26**(4), 232–246 (2007)

# Chapter 6
# Flammability in WPC Composites

## 6.1 Introduction

Wood fiber reinforced plastic composites represent an emerging class of materials that combine the favorable performance and cost attributes of both wood and thermoplastics [1]. In comparison to other fillers, the natural and wood fiber reinforced polymer composites are more environmentally friendly, and are used in transportation, military applications, building and construction industries, packaging, consumer products, etc. [2].

Polypropylene (PP) has been widely used for production of natural fiber/polymer composites because of its low density, high water and chemical resistance, good processability, and high cost–performance ratio [3–6]. Due to the poor compatibility between natural fibers and PP matrix, a compatibilizer can be added to improve adhesion between matrix and fibers leading to enhancement of mechanical properties of composites. A prominent method represents the addition of maleic anhydride polymers as compatibilizers, e.g., maleic anhydride-grafted poly(propylene) (PP-g-MA) and poly(styrene)-*block*poly (ethene-*co*-1-butene)-*block*-poly(styrene) triblock copolymer (SEBS-g-MA) [4–8].

Another drawback of WPCs is their high flammability. As organic materials, the polymers and the wood fibers are very sensitive to flame; improvement of flame retardancy of the composite materials has become more and more important in order to comply with the safety requirements of the wood fiber-composite products [9]. There are two methods for increasing fire retardancy of composites [10]:

- application of different types of fire retardants during the manufacturing process and
- application of fire retardants, especially intumescent coatings, at the stage of finishing.

J. K. Kim and K. Pal, *Recent Advances in the Processing of Wood–Plastic Composites*, Engineering Materials, 32, DOI: 10.1007/978-3-642-14877-4_6,

Fire resistant natural fiber-containing composites can be obtained by the following methods:

- impregnation of natural fibers with fire retardants before the manufacturing process,
- implementation of fire retardants in liquid or solid form during the manufacture of composites,
- application of non-flammable binders, resins, polymers,
- insulation of composites to prevent from penetration of heat flux (intumescent coatings and fire barriers) and
- application of nanoparticles to nature fiber composites.

These methods can be used alternatively or jointly. In general, the addition of non-combustible components is aimed at covering and separating natural fibers or lignocellulosic particles and this is accompanied by migration of fire retardants into the interior of flammable materials, thus resulting in fire protection of the latter. Halogenated flame-retardants, such as organic brominated compounds, are often used to improve the flame-retarding properties of polymers; unfortunately, these also increase both the smoke and carbon monoxide yield rates due to their inefficient combustion [11]. The other commonly used flame retardants boric acid; ammonium phosphates and borates; ammonium sulfate and chloride; zinc chloride and borate; phosphoric acid; dicyanodiamide; sodium borate; and antimony oxide [12, 13]. A considerable attention was paid recently to the application of nano-particles to the modification of polymers, and nanocomposites constitute a new development in the area of flame retardancy and offer significant advantages over conventional formulation where high loadings are often required [14].

There is little research on the flammability of nature fiber and wood fiber composites in the literature. Yap et al. [15] investigated the effects of phosphonates on the flame retarding properties of tropical wood–polymer composites. Anna et al. [16] studied surface treated cellulose fibers in flame retarded PP composites by constituting a high-performance intumescent FR system in the PP matrix, and one of their results showed that the addition of ammonium polyphosphate (APP) to the cellulose fiber containing composite would result in an FR compound. Sain et al. [17] investigated the properties of the composite of PP and chemithermo mechanical pulp reactively treated with bismaleimide-modified PP or premodified pulp. Their results indicated that in situ addition of sodium borate, boric acid, or phenolic resin during processing of the composite decreased the rate of burning of PP. Li and He [18] investigated the flame retardancy and thermal degradation of linear low-density polyethylene (LLDPE)-WF composites. In their study, APP and the mixtures of APP, melamine phosphate (MP) or pentaerythritol (PER) were used as FRs. Their experimental results demonstrated that APP is an effective FR for LLDPE-wood–fiber composites by promoting char formation of the composites. However, the addition of APP (30–40 phr) reduced the Izod impact strength and hardly affected the tensile strength of the composites. Sain et al. [19] found that magnesium hydroxide can effectively reduce the flammability (almost 60%) of natural fiber filled polypropylene composites. No synergetic effect was observed when magnesium hydroxide was used in combination with boric acid and zinc borate, marginal reduction in the mechanical properties of the composites was found with addition of flame-retardants. Zhao et al. [20] reported the mechanical properties, fire retardancy and smoke suppression of the silane-modified WF/PVC composites filled by modified montmorillonite (OMMT), and observed that the fire flame retardancy and smoke suppression of composites were strongly improved with the addition of OMMT. Guo et al. [21]. investigated the effects of nanoclay particles on the flame retarding characteristics of wood–fiber/plastic composites (WPC), the result indicates that using a small amount of nanoclay can significantly improve the flame retarding properties of HDPE/WF nanocomposites.

Ammonium polyphosphate (APP) is an effective intumescent fire retardant for several kinds of polymer-based materials [22–24]. It is high molecular weight chain phosphate. Its efficiency is generally attributed to increase of the char formation through a condensed phase reaction. Silica is usually used as enhancing agent in thermoplastic polymers to increase the mechanical properties, such as tensile strength and toughness. And it has also been recognized as inert diluents and shows some flame retardant effect. Kashiwagi et al. [25] have reported the flame retardant mechanism of silica in polypropylene blends. Fu et al. [26] reported the synergistic flame retardant mechanism of fumed silica in ethylene–vinyl acetate/magnesium hydroxide blends, the results indicated that the addition of a given amount of fumed silica apparently increased the LOI value and decreased the loading of MH in EVA blends. This work is mainly devoted to report the influence of APP and silica on the flammability and thermal decomposition behavior of wood–fiber/PP composite.

In this chapter, the influence of APP and silica on the foamability of wood–fiber/PP composite was investigated by some researchers.

## 6.2 Preparation Wood–Fiber/PP Composites by Twin Screw Extruder

All samples showed in Table 6.1 were prepared by co-rotating intermeshing twin-screw extruder (Bau-Tech, Korea). It has a screw diameter of 19 mm and the distance between screw axes is 18.4 mm with L/D ratio of 40. It is fitted with a modular screw configuration, which has different combinations of right-handed and left-handed screws and neutral kneading disc elements with one reverse-pumping screw elements as shown in Fig. 6.1. The screw speed was fixed 160 rpm while the cylinder temperature was maintained at 160, 166, 176, and 180°C from the hopper to the die. The extrudate was pelletized and dried under vacuum at 80°C for 24 h to remove any residual water. At last the samples were molded at temperature profile of 160/170/170/180°C by injection for mechanical testing.

**Table 6.1** Formulation of wood–fiber/PP Composites and LOI results

| Samples[a] | APP | Silica | LOI |
|---|---|---|---|
| a | 0 | 0 | 21.4 |
| b | 10 | 0 | 24.5 |
| c | 20 | 0 | 26.5 |
| d | 30 | 0 | 27.4 |
| e | 40 | 0 | 27.9 |
| f | 20 | 2 | 27.1 |
| g | 20 | 6 | 28.4 |
| h | 20 | 10 | 28.9 |

*APP* ammonium polyphosphate, *LOI* limiting oxygen index
[a] Base material is PP 65 phr; wood–fiber 30 phr; PP-g-MA 5 phr, SEBS-g-MA 5 phr

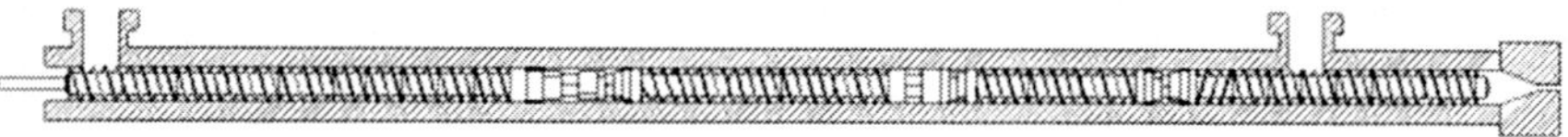

**Fig. 6.1** Screw configuration used for the PP/WGRT blending experiments

## 6.3 Preparation and Analysis of Wood–Fiber/PP Composite Foams

The wood–fiber/PP composite after extrusion was pelletized; the plate samples of the composites with 2.0 mm thickness were compression-molding at 180°C for 6 min.

Microcellular foaming experiments were performed in a batch process. A schematic of the batch-foaming process is shown in Fig. 6.2. Plate samples that were 2.0 mm thick, 60 mm long, and 4.0 mm wide were enclosed high-pressure vessel. The vessel was flushed with low-pressure $CO_2$ for about 3 min and pressurized to the saturated vapor pressure $CO_2$ at room temperature and preheated to desired temperature. Afterward, the pressure was increased to the desired pressure by a syringe pump and maintained at this pressure for 2 h to ensure equilibrium absorption of $CO_2$ by the samples. After saturation, the pressure was quenched atmospheric pressure within 3 s and the samples were taken out. Then foam structure was allowed to full growth during rapid depressurization.

The density of foam and unfoamed samples was determined from the sample weight in air and water, respectively, according to ASTM D 792 method A. Then the density of the foamed sample is divided by the density of the unfoamed sample to obtain the relative density ($\rho_r$).

$$\rho_r = \frac{\rho_f}{\rho_m} \tag{6.1}$$

where $\rho_m$ and $\rho_f$ are the density of the unfoamed polymer and foamed polymer, respectively.

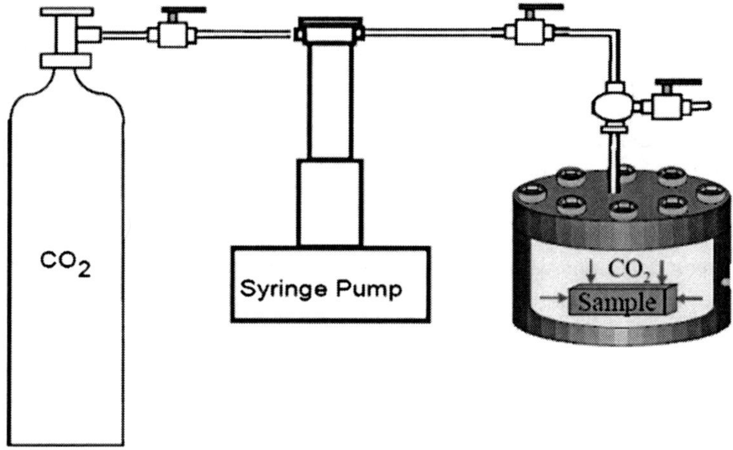


**Fig. 6.2** The schematics of batch-foaming process

## 6.4 Mechanical Properties of PP/Wood Fiber Composites

The mechanical properties of the wood–fiber/PP composites are shown in Figs. 6.3 and 6.5. Despite the presence of the compatibilizer, the addition of APP as flame-retardant shows the decrease trend of tensile strength and elongation at break. This could be attributed to the poor compatibility of the added flame retardant with polymer. The next important point, causing such a decrease, is the existence of the cavities within the samples, formed via thermal decomposition of fillers and release of steam during the process. Deterioration of the mechanical properties of the filled and unfilled plastics with the addition of flame-retardants has reported by some workers [27, 28].

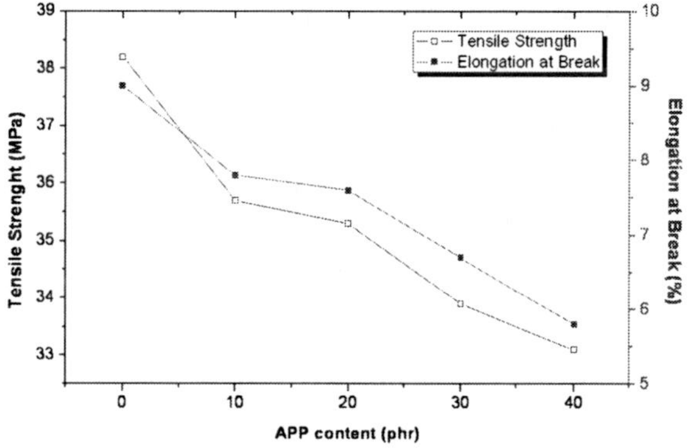


**Fig. 6.3** The effect of APP content on tensile strength and elongation at break of wood–fiber/PP composite

Silica is usually used as an enhancing agent in thermoplastic polymers to increase the mechanical properties, such as tensile and toughness. But a higher loading level of inorganic fillers or flame retardants would lead to earlier breaking of the composites. Thus, the tensile strength of the composites increase with the addition of silica, however, obvious decrease can be observed with more than 4 phr silica. This attribute to the larger size of silica agglomerates at higher amount of silica. These agglomerates can act as stress concentrators and mechanical failure points initiate the fracture of the specimens [29]. The elongation at break decreases with increasing of silica content.

The increase in amount of filler (APP or silica) causes a decrease in impact strength (Figs. 6.4, 6.5, 6.6), which can be correlated to fragile surface adhesiveness, among fillers, matrix and cavities formed within the sample. Typically, a polymer matrix with high loading of fillers has less ability to absorb impact energy, because the filler disturbs matrix continuity and each particle is a site of stress concentration, which can act as micro-crack initiator [30].

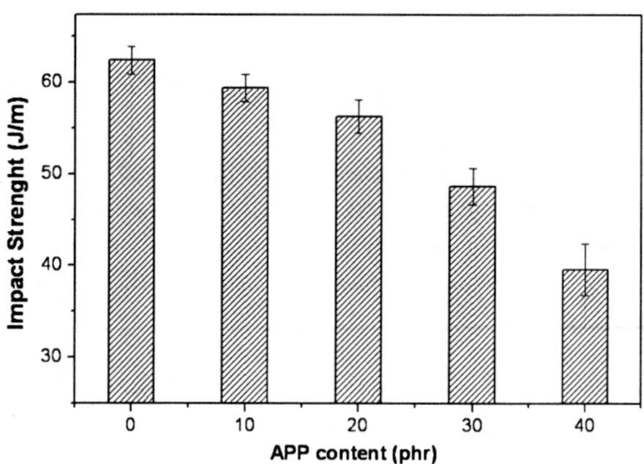


**Fig. 6.4** The effect of APP content on impact strength of wood–fiber/PP composite

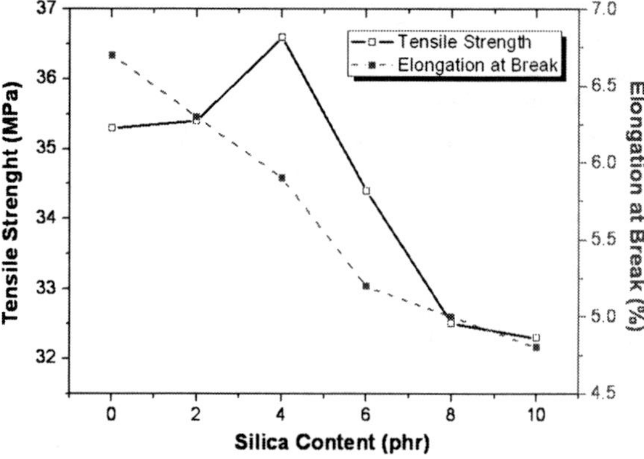


**Fig. 6.5** Effect of silica content on the tensile strength and elongation at break of wood–fiber/PP/APP composites

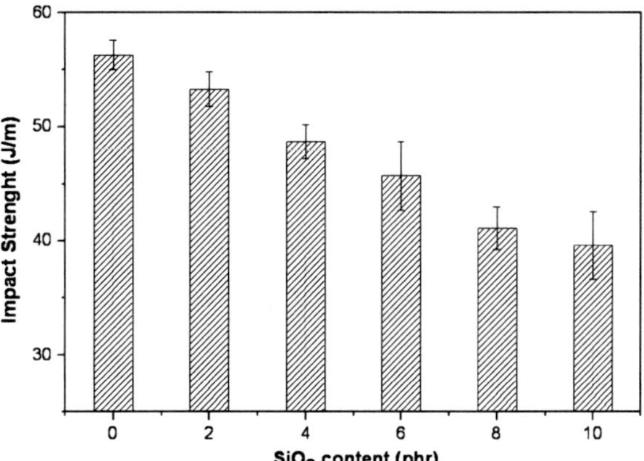


**Fig. 6.6** Effect of silica content on the impact strength of wood–fiber/PP/APP composites

## 6.5 Flame Retardancy of PP/Wood Fiber Composites

Limited oxygen index (LOI), a widely used method as a simple and precise method for the determination of fire self-extinguishment, was adopted to evaluate the flame retardant properties of wood–fiber/PP/APP/silica composites. Table 6.1 gives LOI data of all wood–fiber/PP composites samples. From the experimental results shown in Table 6.1, wood–fiber/PP composite is easy flammable and its LOI is only 21 because PP and wood are easily flammable materials. APP is an important flame retardant for wood cellulose and an important component of intumescent flame retardants. Which obviously enhance flame retardancy of wood–fiber/PP composite, who's LOI reached 27.9 when addition of APP is 40 phr. So it can be concluded that APP is a very effective flame retardant for wood–fiber/PP composite, because APP can effectively catalyse wood or natural fiber to form char. From Table 6.1, it also can be seen that when the APP content is fixed at 20 phr, the LOI increases with increasing of silica content, for example, the LOI value of sample d with 30 phr APP is only 27.4%, whereas the sample h, APP and silica is 20 and 10 phr in the composite, respectively, increased to 28.9%. Apparently, there is synergistic effect of silica and APP on enhancing the LOI of wood–fiber/PP composite.

## 6.6 Thermal Degradation of PP/Wood Fiber Composites

The results of the thermal analysis are shown in Figs. 6.7 and 6.8. All materials decomposed in two decomposition steps. Table 6.2 summarizes the characteristics for the two subsequent decomposition steps. For all the materials, the first step is attributed to the decomposition of wood–fiber, the second one to the decomposition

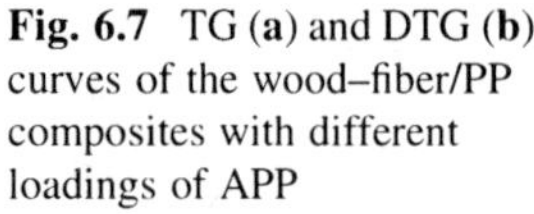

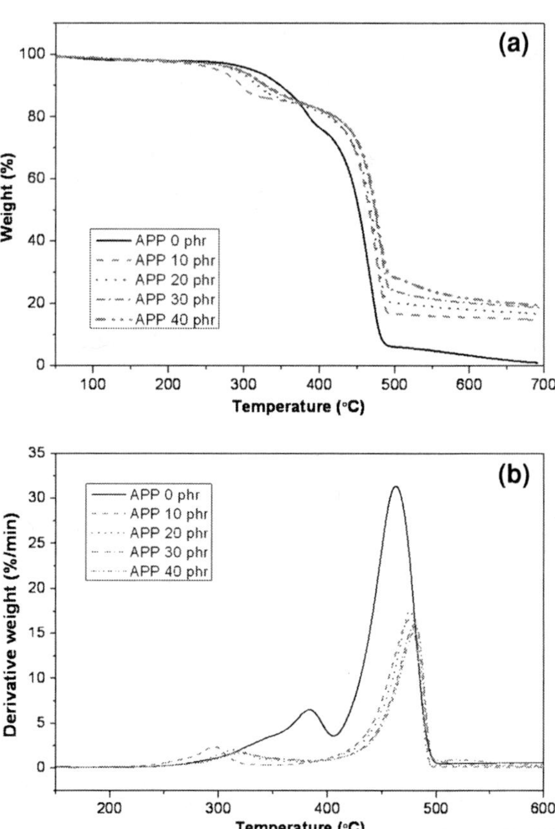


**Fig. 6.7** TG (**a**) and DTG (**b**) curves of the wood–fiber/PP composites with different loadings of APP

of the PP and compatibilizer. This assignment considers the results for the distinct steps with respect to mass loss and the temperature interval. With addition of APP to the wood composite accelerates the first step of decomposition. For example, adding 10 phr APP, the temperature for the maximum mass loss is around 81°C below the temperature without APP. Hence, the temperature interval for the first decomposition step does not only correspond to the decomposition of the wood–fiber, but is also typical for the release of $NH_3$ and $H_2O$ from the APP [31]. The char residual in part from the wood–fiber, and in part is composed of the APP-based additive. The latter is expected to show heat insulation properties, so the second decomposition is shifted to higher temperature as expected. Similar effect was observed for APP on other materials such as polyamide/ethylene–vinyl-acetate, polyurethane, and PP/flax blends [32–34].

All the above results indicate that the addition of APP to the wood–fiber/PP composite decreases the thermal stability of wood–fiber/PP composite at first stage and increases the thermal stability at the second stage. Therefore, the thermal degradation of wood–fiber/PP composite may take place as follows: at the first

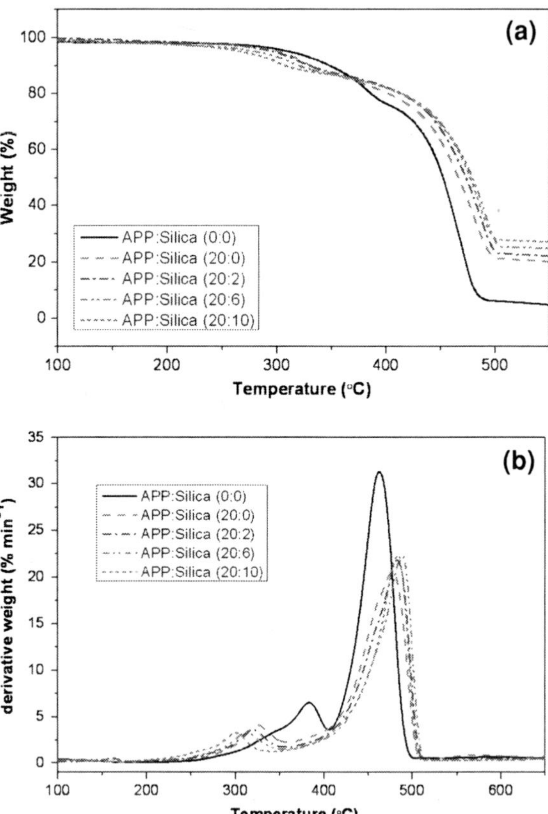


**Fig. 6.8** TG (**a**) and DTG (**b**) curves of the wood–fiber/PP composites with different loadings of silica

**Table 6.2** TG results of the composites

| Sample | $T_{DTGmax}$ 1st step (°C) | Mass loss 1st step (%) | $T_{DTGmax}$ 2nd step (°C) | Mass loss 2nd step (%) | Char residue 600°C (%) |
|---|---|---|---|---|---|
| a | 383.39 | 81.81 | 462.34 | 34.64 | 2.96 |
| b | 302.06 | 89.99 | 474.13 | 36.86 | 15.72 |
| c | 317.75 | 90.85 | 475.31 | 38.03 | 18.16 |
| d | 322.65 | 91.25 | 480.03 | 39.70 | 20.35 |
| e | 322.65 | 91.48 | 478.85 | 43.99 | 21.69 |
| f | 314.06 | 91.81 | 483.57 | 38.89 | 21.05 |
| g | 308.17 | 91.31 | 488.51 | 39.69 | 24.23 |
| h | 298.78 | 91.13 | 484.16 | 41.87 | 26.46 |

stage, APP interacts with wood composite, leading to generation of volatile compounds and a phosphorus rich layer, which could protect the polymer matrix under heat, and then the protective layer would decompose to yield a compact char on the surface of the materials to protect the polymer matrix effectively at the second stage.

Addition silica also accelerates the first step of decomposition, and generates more char residual. The temperature of the second decomposition is shifted to higher temperature. These shifts are due to the silica incorporates with APP to form a charred layer and inhibits the heat and mass transfer between surface and melting polymer, resulting in the increase of fire resistance of the composites.

## 6.7 Cone Calorimeter Study of PP/Wood Fiber Composites

The cone calorimeter based on the oxygen consumption principle has widely been used to evaluate the flammability characteristics of materials. Although a cone calorimeter test is in a small-scale, the obtained results have been found to correlate well with those obtained from a large-scale fire test and can be used to predict the combustion behavior of materials in a real fire.

The HRR measured by cone calorimeter is very important parameter as it expresses the intensity of a fire. A highly flame retardant system normally shows a low av-HRR value. The p*k*-HRR value is used to express the intensity of a fire. The changes of HRR as a function of burning time for different samples a, c, d, and h is shown in Fig. 6.9.

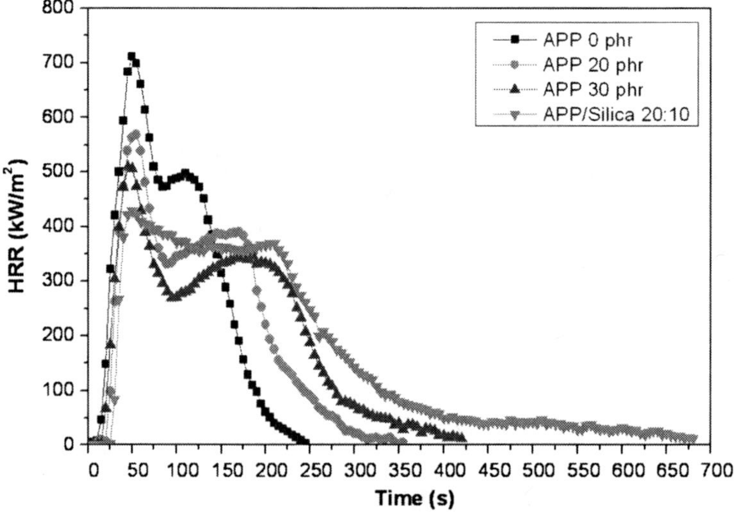


**Fig. 6.9** HRRs versus burning time for different wood–fiber/PP composites

It can be found from Fig. 6.9 that PP/WF composite (sample a) burns fast after ignition, and have a sharp HRR curve at range of 40-80 s, whereas sample c and d with 20 and 30 phr APP, respectively, show a decline of the HRR curve, the combustion of sample c and d is prolonged to 360 and 470 s, respectively, from 260 s of the control sample a. Silica is usually considered to be an inert additive in flame retardant polymers. It can be found that sample h containing 20 phr APP and 10 phr silica shows a decline of the HRR curve compare with the sample d which containing 30 phr APP, the burning was also prolonged to 670 s. Table 6.3 lists the data of TTI, HRR, and MLR obtained from the cone calorimeter tests of sample a, c, d, and h. It can be found that the values HRR and MLR decrease by addition of APP and silica, which also prolonged the TTI. Sample h with 10 phr silica and 20 phr APP shows the lower HRR and MLR than sample d with 30 phr APP. This result further gives the evidence that the 10 phr silica has the synergist effect with APP in the wood–fiber/PP composites.

The above data indicate that the fire-resistance performance of wood–fiber/PP composite is enhanced by partly substituting APP with silica. This is mainly due to the silica tends to accumulate on the surface in fire and consequently forms a charred layer by combining with APP. Another reason is that the silica can prevent the cracking of the char layer. Wrinkles and cracks are found on the char layer formed during the combustion of wood–fiber/PP composites as shown in Fig. 6.10a, b, however, it is observed that smoother and more compact char is formed with addition of silica as shown in Fig. 6.10c. This charred layer prevented heat transfer and transportation of degraded products between melting polymer and surface, thus reduced the HRR and MLR.

## 6.8 SEM Morphological Observation

The mechanical properties are influenced by the shape, dimension, and dimension distribution of the filler particles, the filler content, and the physical properties of the filler and polymer, the interfacial interaction between the polymer and the filler is one of the most important factors. The interfacial interaction between the polymer and filler not only can change the local deformation and micro mechanism of the local deformation and the breaking process, but also can influence the crystalline behavior of the polymer, both of which influence the mechanical properties of the composites [35]. Morphologies of the APP and silica filled wood–fiber/PP composites are explored by SEM. Figure 6.11(A) are SEM micrographs of impact fracture sections of composite (a), (c) (e) and (h). With addition of APP and silica, as shown in Fig. 6.11(A), in some domains, larger particles of agglomerated APP filler can be observed; these larger particles lead to a deterioration of the mechanical properties.

**Table 6.3** Cone data of some wood–fiber/PP composite samples

| Sample | APP (0 phr) Sample a | APP (20 phr) Sample c | APP (30 phr) Sample d | APP + silica (20 + 10 phr) Sample h |
|---|---|---|---|---|
| TTI (s) | 12 | 18 | 18 | 32 |
| Pk-HRR (kW $m^{-2}$) | 701 | 568 | 505 | 428 |
| Av-HRR (kW $m^{-2}$) | 301 | 215 | 199 | 156 |
| Pk-MLR (g $s^{-1}$) | 0.175 | 0.145 | 0.124 | 0.11 |

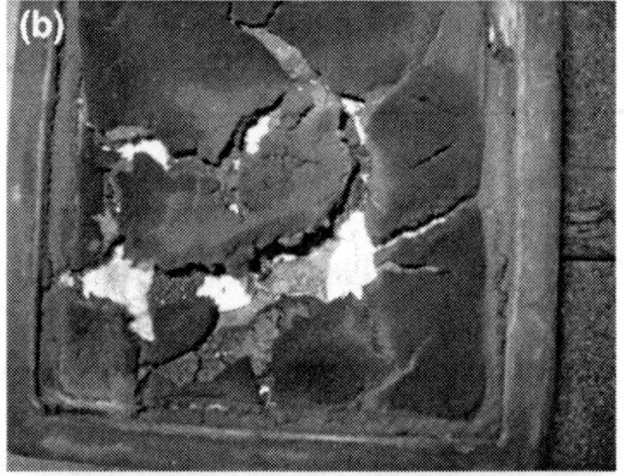

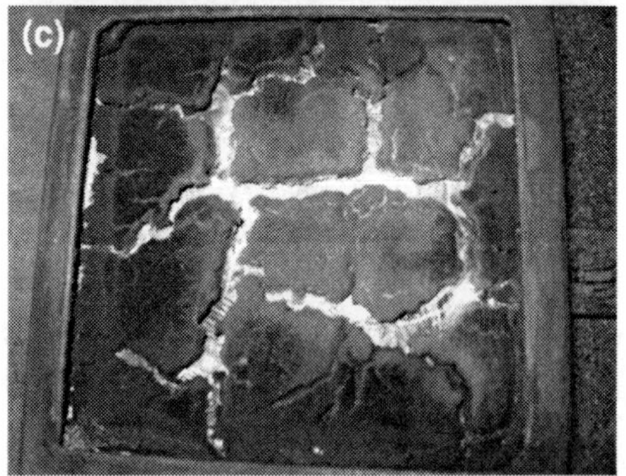


**Fig. 6.10** Char formation for some different wood–fiber/PP composite samples. **a** Sample c, **b** sample d, **c** sample h

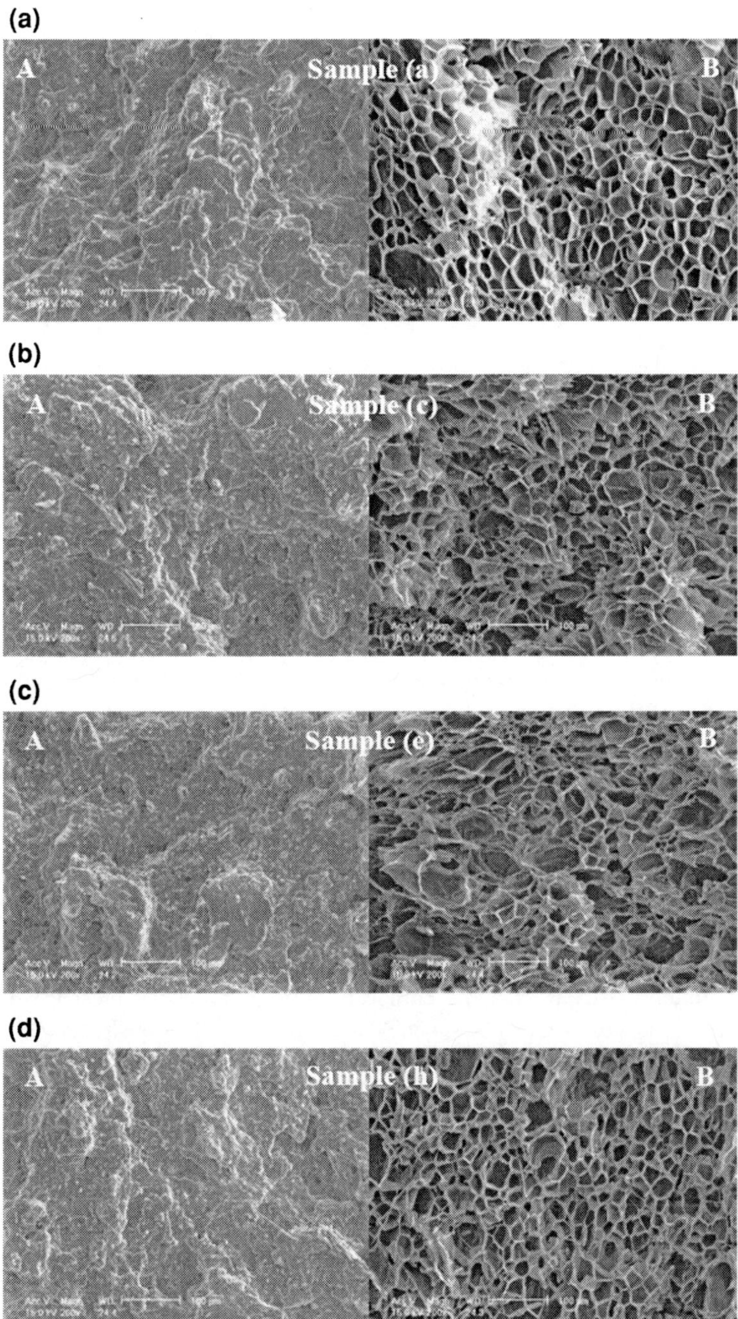


**Fig. 6.11** FESEM microphotographs of the dispersion of flame retardants (*A*) and microcellular foam (*B*) (24 MPa, 160°C) of **a** sample-a, **b** sample-c, **c** sample-e and **d** sample-h which are listed in Table 6.1

## 6.9 Batch Physical Foaming

### *6.9.1 Effect of APP and Silica*

Figure 6.11a–d shows the SEM micrographs of the wood–fiber/PP composite, APP filled and APP-silica filled wood–fiber/PP composites. As can be seen in Fig. 6.11, the cell shape of composite (a) is closed cellular polyhedron, no collapse or collision in the cell system is observed. The composite (c) and (e) are wood–fiber/PP composites filled with 20, 40 phr APP, respectively. From the microphotographs of composite (c) and (e), it can be observed that with addition of APP, some cell size become large and irregular shape compared with pure wood–fiber/PP composite, the cells become non-spherical and non-uniform. This is because with addition of APP results in higher viscosity, so it is more difficult to disperse the APP particles in wood–fiber/PP composites and lead to APP agglomerates, and it destroys the closed regular cellular polyhedron.

From the Fig. 6.11, the micrograph of composite (h) is quite different from the only APP filled with wood–fiber/PP composite, the number of big cells decreased, and less cell collapse or collision in the cell system was observed, the cell structure become more uniform. This result can be explained as follows: the silica substitute for a part of APP particle, thus reducing the breakage of closed regular cell, the silica isolates the APP particle at a certain extent, which can also reduce the breakage of cell.

Figure 6.13 and 6.15 illustrate the effect of the APP and silica loading on the relative density of wood–fiber/PP composite, it can be observed that the relative density slightly decreased, while increasing the content of fillers, and then relative density increased. As aforementioned, because of the APP agglomerates and the comparatively big caused the some large cell size and thinner cell wall, which resulted to the lower relative density. However, the addition of too much content of fillers to polymer matrix leads to a poor surface adhesion between filler and polymer matrix, and the poor surface adhesion provides a channel through which gas can quickly escape from the composites. In addition, the presence of filler increases the viscosity of wood–fiber/PP composites, which obstructs cell nucleation and growth, so the relative density increased.

### *6.9.2 Effect of Saturation Pressure*

The effect of pressure on the final cells structure is studied at constant temperature (160°C) and depressurization time of 3 s, while pressure ranged between 8 and 24 MPa. Typical cells structures obtained are presented in Fig. 6.12. As shown in Figs. 6.12 and 6.13, the cell size increases, while relative density decreases by increasing the pressure. Increasing saturation pressure, the extent of the $CO_2$ induced-melting temperature depression increases, namely the amount of $CO_2$

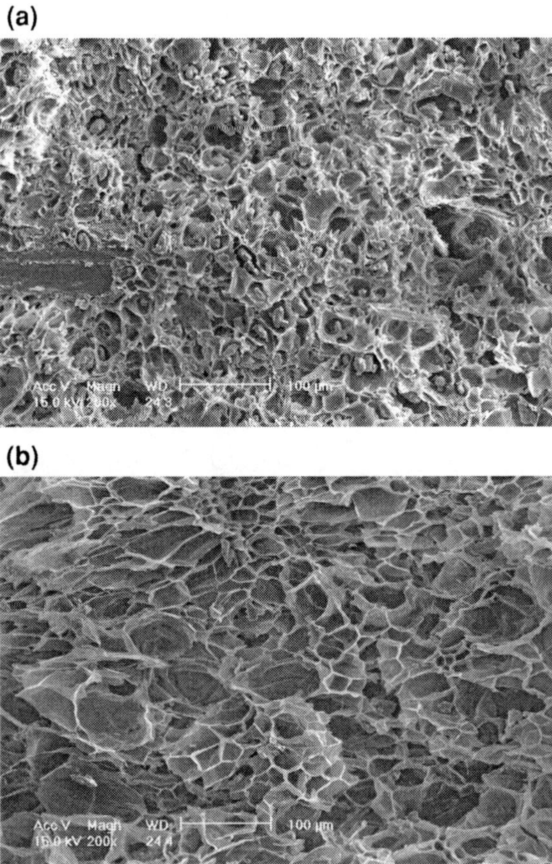


**Fig. 6.12** Typical cell structures of sample (e) composite foams at 160°C and **a** 112 MPa, **b** 24 MPa

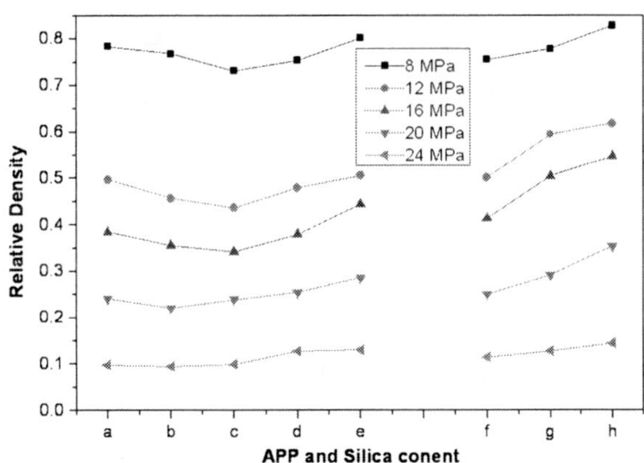


**Fig. 6.13** Effect of pressure on the relative density of wood–fiber/PP composites

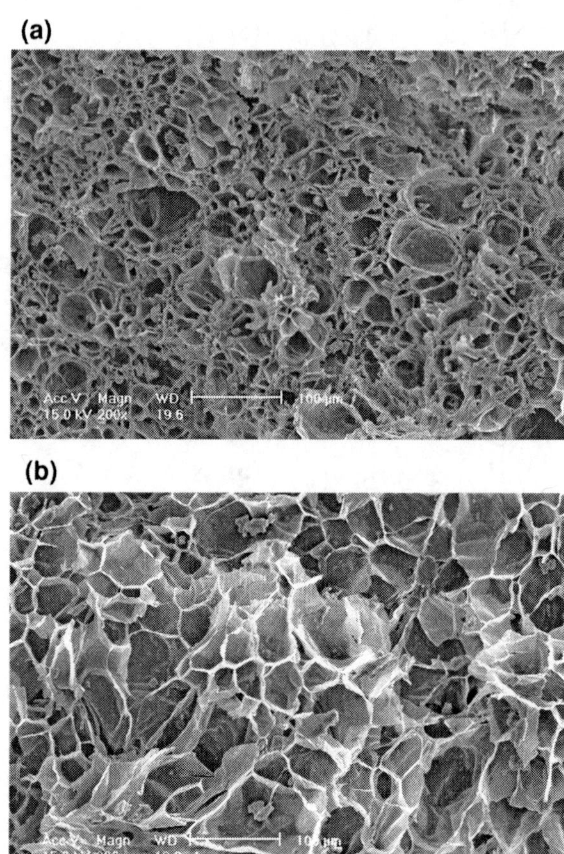


**Fig. 6.14** Typical cell structures of composite-e foams at 20 MPa and **a** 146°C, **b** 166°C

dissolved in samples increases. As a result, the melt strength of the matrix is weaker, so the sample is relatively soft and its deformability is large. On the other hand, at the low saturation pressure, the depressurization rate is low, the level of supersaturation is low and the driving force for the cell nucleation is weak, so the cell size is smaller at low pressure. So the result is that increasing saturation pressure becomes more favorable for sample to foam and for the cells to grow bigger in size, the larger cell sizes and the thinner cell walls brings on the reduced densities and relative densities [36].

### *6.9.3 Effect of Saturation Temperature*

The effect of temperature on the final cell structure was studied at constant pressure of 20 MPa. Temperature varied between 136 and 166°C. Typical cell structures are presented in Fig. 6.14. As shown in Figs. 6.14 and 6.15, an

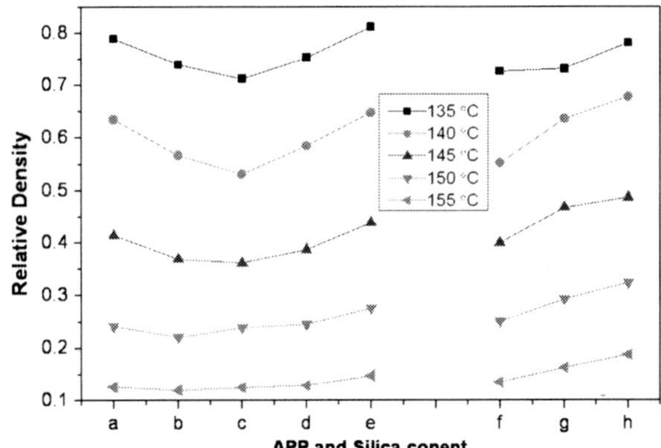


**Fig. 6.15** Effect of temperature on the relative density of composites

increase of temperature lead to a significant increase in the average cell size, while it led to a decrease in the relative density. This behavior is typical for the foaming of many polymers with $CO_2$ [37, 38]. The cell size reflects the effect of many factors such as the number of the formed nuclei, the amount of the dissolved gas, and its diffusivity, and the viscosity of the polymer matrix. At higher temperatures the energy barrier to nucleation decreases as predicted from nucleation theory [37]. The generation of nuclei becomes more difficult and, as consequence, fewer cells are observed in the final cell structure (reduced cell density). Simultaneously, the $CO_2$ solubility in the polymer matrix decreases and, consequently, there is less fluid available for nucleation and growth of pores. Furthermore, as temperature increases, the viscosity of the polymer matrix decreases facilitating the grower and also the coalescence of neighboring cells. Also, the diffusivity of the fluid increases resulting in the faster growth of cells. Facilitated and faster cell growth leads to formation of lager cells and foams with reduced relative density.

## References


1. Faruk, O., Bledzki, A.K., Matuana, L.M.: Microcellular foamed wood–plastic composites by different process: a review. Macromol. Mater. Eng. **192**, 113–127 (2007)
2. Kozlowski, R., Wladyka-Przybylak, M.: Uses of natural fiber reinforced plastics. In: Wallenberger, F.T., Weston, N.E. (eds.) Natural Fibers, Plastics and Composites. Kluwer, Dordrecht (2004)
3. Ichazo, M.N., Albano, C., Gonzalez, J., Perera, R., Candal, M.V.: Polypropylene/wood flour composites: treatments and properties. Compos. Struct. **64**(2), 207–214 (2001)
4. Mohanty, S., Nayak, S.K., Verma, S.K., Tripathy, S.S.: Effect of MAPP as a coupling agent on the performance of jute-PP composites. J. Reinforc. Plastics Compos. **23**, 626–637 (2004)
5. Feng, D., Caulfield, D.F., Sanadi, A.R.: Effect of compatibilizer on the structure–property relationships of kenaf-fiber/polypropylene composites. Polym. Compos. **22**, 606–617 (2001)

6. Bledzki, A.K., Faruk, O.: Wood fibre reinforced polypropylene composites: effect of fibre geometry and coupling agent on physico-mechanical properties. Appl. Compos. Mater. **10**(6), 366–379 (2003)
7. Pracella, M., Chionna, D., Anguillesi, I., Kulinski, Z., Piorkowska, E.: Functionalization, compatibilization and properties of polypropylene composites with Hemp fibres. Compos. Sci. Technol. **66**(13), 2218–2230 (2006)
8. Wu, J., Yu, D., Chan, C.M., Kim, J., Mai, Y.W.: Effect of fiber pretreatment condition on the interfacial strength and mechanical properties of wood fiber/PP composites. J. Appl. Polym. Sci. **76**(7), 1000–1010 (2000)
9. Sain, M., Park, S.H., Suhara, F., Law, S.: Flame retardant and mechanical properties of nature fiber-PP composites containing magnesium hydroxide. Polym. Degrad. Stab. **83**, 362–367 (2004)
10. Kozlowski, R., Helwig, M.: Progress in fire retardants for lignocellulosic materials. In: Proceedings of the 6th Arab international conference on materials science, materials & fire, Alexandria, Egypt, pp. 1–11 (1998)
11. Pape, P.G., Romenesko, D.J.: The role of silicone powders in reducing the heat release rate and evolution of smoke in flame retardant thermoplastics. J. Vinyl Addit. Technol. **3**, 226–232 (1997)
12. Lewin, M., Atlas, S.M., Pearce E.M.: Flame-Retardant Polymeric Materials. Plenum Press, New York (1976)
13. Kozlowski, R., Wladyka-Przybylak M.: Natural polymers: wood and lignocellulosics. In: Horrocks, R. (ed.) Fire Retardant Materials. Woodhead Publishing Limited, Cambridge (2000)
14. Kandola, B.: Nanocomposites. In: Horrocks, A.R. (ed.) Fire Retardant Materials. Woodhead Publishing Limited, Cambridge (2000)
15. Yap, M.G.S., Que, Y.T., Chia, L.H.L., Chan, H.S.O.: Thermal properties of tropical wood-polymer composites. J. Appl. Polym. Sci. **43**, 2057–2065 (1991)
16. Anna, P., Zimonyi, E., Mariton, A., Matko, S., Keszi, S., Bertalan, G., Marosi, G.: Surface treated cellulose fibres in flame retarded PP composites. Macromol. Symp. **202**, 246–264 (2003)
17. Sain, M.M., Kokta, B.V.: Polyolefin-wood filler composite. I. Performance of *m*-phenylene bismaleimide-modified wood fiber in polypropylene composite. J. Appl. Polym. Sci. **64**, 1646–1669 (1994)
18. Li, B., He, J.: Investigation the mechanical property, flame retardancy and thermal degradation of LLDPE-wood fiber composites. Polym. Degrad. Stab. **83**, 241–246 (2004)
19. Sain, M., Park, S.H., Suhara, F., Law, S.: Flame retardant and mechanical properties of natural fiber-PP composites containing magnesium hydroxide. Polym. Degrad. Stab. **83**(2), 363–367 (2004)
20. Zhao, Y., Wang, K., Zhu, F., Xue, P., Jia, M.: Properties of poly(vinyl chloride)/wood flour/montmorillonite composites: effects of coupling agents and layered silicate. Polym. Degrad. Stab. **91**(2), 2874–2883 (2006)
21. Guo, G., Park, C.B., Lee, Y.H., Kim, Y.S., Sain, M.: Flame retarding effects of nanoclay on wood–fiber composites. Polym. Eng. Sci. **47**, 330–336 (2007)
22. Matko, S., Toldy, A., Keszei, S., Anna, P., Bertalan, Gy., Marosi, Gy.: Flame retardancy of biodegradable polymers and biocomposties. Polym. Degrad. Stab. **88**, 138–146 (2004)
23. Duquesne, S., Bras, M.L., et al.: Mechanism of fire retardancy of polyurethanes using ammonium polyphosphate. J. Appl. Polym. Sci. **82**, 3262–3274 (2001)
24. Balabanovich, A.I.: The effect of ammonium polyphosphate on the combusition and thermal decomposition behavior of poly(butylenes terephthalate). J. Fire Sci. **21**, 286–298 (2003)
25. Kashiwagi, T., Gilman, J.M., Butler, K.M., Harris, R.H., Shields, J.R., Asano, A.: Flame retardant mechanism of silica gel/silica. Fire Mater. **24**, 277–289 (2000)
26. Fu, M., Qu, B.: Synergistic flame retardant mechanism of fued silica in ethylene–vinyl acetate/magnesium hydroxide blends. Polym. Degrad. Stab. **86**, 633–639 (2004)

27. Sain, M., Park, S.H., Suhara, F., Law, S.: Flame retardant and mechanical properties of natural fibre-PP composites containing magnesium hydroxide. Polym. Degrad. Stab. **83**, 363–367 (2004)
28. Chiu, S.H., Wang, W.K.: The dynamic flammability and toxicity of magnesium hydroxide filled intumescent fire retardant polypropylene. J. Appl. Polym. Sci. **67**, 989–996 (1998)
29. Pavlidou, E., Bikiaris, D., Vassiliou, A., Chiotelli, M., Karayammidis, G.: Mechanical properties and morphological examination of isotactic polypropylene/$SiO_2$ nanocomposites containing PP-g-MA as compatibilizer. J. Phys. Conf. Ser. **10**, 190–193 (2006)
30. Mareri, P., Bastide, S., Binda, N., Crespy, A.: Mechanical behaviour of polypropylene composites containing fine mineral filler: effect of filler surface treatment. Compos. Sci. Technol. **68**, 747–762 (1998)
31. Statheropoulos, H., Kyriakou, S.A.: Quantitative thermogravimetric-mass spectrometric analysis for monitoring the effects of fire retardants on cellulose pyrolysis. Anal. Chim. Acta **409**, 203–214 (2000)
32. Siat, C., Bourbigot, S., Le Bras, M.: Thermal behavior of polyamide6-based intumescent formulations—a kinetic study. Polym. Degrad. Stab. **68**, 303–313 (1997)
33. Meng, X.Y., Ye, L., Zhang, X.G., Tang, P.M., Tang, J.H., Xu, J., Li, Z.-M.: Effects of expandable graphite and ammonium polyphosphate on the flame-retardant and mechanical properties of rigid polyurethane foams. J. Appl. Polym. Sci. **114**, 863 (2009)
34. Schartel, B., Braun, U., Schwarz, U., Reinemann, S.: Fire retardancy of polypropylene/flax blends. Polymer **44**, 6241–6260 (2003)
35. Lu, M., Zhang, S., Yu, D.: Study on poly(propylene)/ammonium polyphosphate composites modified by ethylene-1-octene copolymer grafted with glycidyl methacrylate. J. Appl. Polym. Sci., **93**, 412–419 (2004)
36. Xu, Z.M., Jiang, X.L., Liu, T., Hu, G.H., Zhao, L., Zhu, Z.N., Yuan, W.K.: Foaming of polypropylene with supercritical carbon dioxide. J. Supercrit. Fluids **41**, 299–310 (2007)
37. Goel, S.K., Beckman, E.J.: Generation of microcellular polymeric foams using supercritical carbon dioxide. I: effect of pressure and temperature on nucleation. Polym. Eng. Sci. **34**, 1137–1147 (1994)
38. Arora, K.A., Lesser, A.J., Mccarthy, J.T.: Preparation and characterization of microcellular polystyrene foams processed in supercritical carbon dioxide. Macromolecules **31**, 4614–4620 (1998)

# Chapter 7
# Expanded Wood Polymer Composites

## 7.1 Introduction

Foamed solid polymers, also referred as microcellular composites or expanded composites or sponge materials are a class of materials that are extensively used in everyday applications from the foamed polyurethane mattress we sleep onto the polystyrene based Styrofoam cup in which we have our morning coffee. According to some estimates, the market of foam products in the world stands at 14 billion US Dollars and is expected to grow at a phenomenal rate of 14% for the next 5 years. This growth will be largely triggered by phenomenal demand of China who in recent times accounts for 26% of world's polystyrene and 34% of polyurethane foam world's consumption. Several polymers can be foamed to desired low densities to suit applications based on properties such as weight reduction, insulation, buoyancy, energy dissipation, convenience and comfort characteristics like cushioning factor etc. A closed loop decision process of polymeric foam performance dependency summary chart is shown in Fig. 7.1.

Based on morphology and physical characteristics, foams can be broadly classified into two main categories:

- Open cell foams, the so called sponges
- Closed celled foams.

Open cell foams are mainly used in low end consumer applications like mattresses and sponges and acoustic insulation whereas closed cell foams can be used as engineering materials and can be used as structural components in construction and automotive applications.

According to the processing methods, plastic foams can be classified as

- Thermoplastic foams
- Thermoset foams.

J. K. Kim and K. Pal, *Recent Advances in the Processing of Wood–Plastic Composites*, Engineering Materials, 32, DOI: 10.1007/978-3-642-14877-4_7,

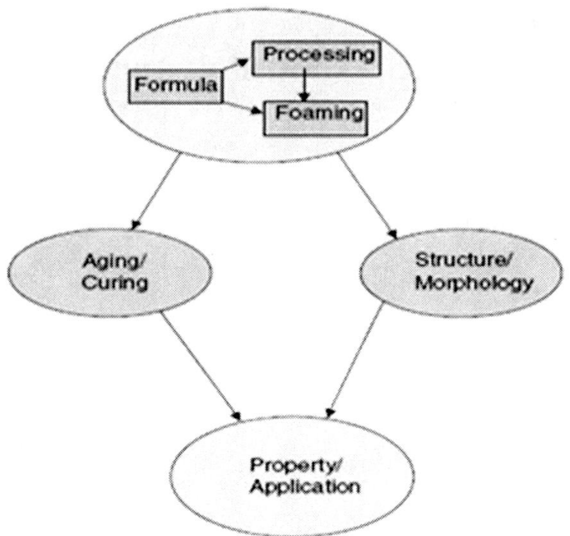


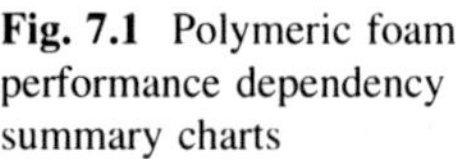
**Fig. 7.1** Polymeric foam performance dependency summary charts

Thermoplastic foams form a major share of microcellular foam market since they can be recycled and reprocessed unlike thermoset foams which are usually characterized by substantial crosslinking and can be reprocessed.

Depending on the physical and mechanical properties, polymer foams can also be classified as rigid, flexible or semi-flexible microcellular foams. The degree of crystallinity, extent of reinforcements and glass transition temperature has a major impact in this type of classification. The three main commercial resins used in foaming applications are: polyurethane, polystyrene and polyvinyl chloride and are mainly used in automotive, construction, packaging and consumer products.

All plastic foams irrespective of their classification consist of two phases, a solid phase made up of polymer and reinforcements and gas phase made up of blowing agent. There are two major foaming methodologies in polymer foam industry: soluble foaming (physical foaming methods like super critical blowing) and reactive foaming (chemical blowing agents). In the physical method typically an inert gas such as carbon dioxide, nitrogen or volatile hydrocarbons such as propane or isopentane are used as blowing agents. In case of chemical blowing agents, compounds that are capable of releasing a gas either nitrogen or carbon dioxide is used to generate the gas phase during processing. Typically azocompounds like ADC (azo di carbamide) are used. In some cases like in food foams (such as cakes), salts which are capable of releasing gases or used (baking soda which releases carbon dioxide during baking). Even alcohols have also been used as blowing agents in certain kinds of food foams.

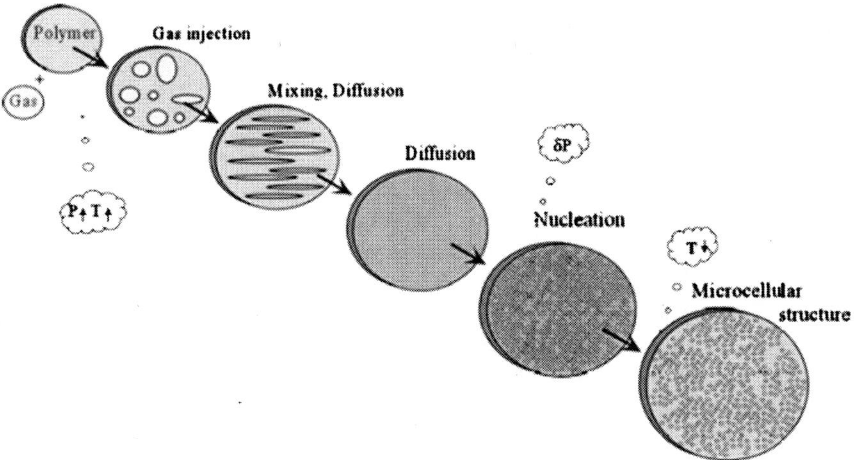


**Fig. 7.2** Various stages involved in production of microcellular composites. Terms *P* and *T* denote pressure and temperature, $\delta P$ denotes reduction in pressure

The basic steps involved in microcellular production are shown in the Fig. 7.2.

The next section gives a brief description of various foaming processes (chemical blowing and super critical foaming) of polymer composites.

## 7.2 Chemical Blowing Agents

Chemical blowing agents (CBA) are the most easy and versatile ways of producing microcellular composites. The main advantage of CBA method of foam processing is that no extra modifications to existing polymer processing equipment are required. CBA can be just compounded and added within the polymer matrix in the hopper of extrusion and injection molding machines. However fine tuning of the temperature profile may be required in order to prevent premature decomposition and to obtain a uniform distribution of the cells within the polymer matrix. However sufficient care must be taken to see that thermal decomposition occurs after the polymer is melted. Once melted the polymer is extruded into a mold maintained at a higher temperature where the CBA is decomposed into gas. Chemical blowing agents or CBAs are a class of materials that decompose to form gases under processing conditions. Most CBAs are solids and decompose within a given temperature range. The various requirements for a chemical to be CBA are:

- Decomposition temperature of the CBA should be slightly higher than the melt or processing temperature of the material.
- Gas must be liberated within a narrow range of temperature.
- Rate of liberation of gas must be sufficient and controllable.
- Gas and the decomposition products must be nontoxic, noncorrosive and nonflammable.

- CBA should have good dispersion in the polymer matrix.
- Since most CBA decompose by exothermic reactions, the heat generated during decomposition should not degrade the polymer matrix.
- Internal diffusion and gas pressure should be controlled.
- The gases generated by CBA should have high diffusion coefficients in the chosen polymer matrix.

Depending on the gas produced by the CBAs, a wide range of cellular polymers can be obtained. For example carbon dioxide releasing CBA normally produce low density foams since $CO_2$ is more soluble in thermoplastics rather than nitrogen. (Henry's law's constant is 2.5–4 times greater in $CO_2$ when compared to $N_2$ in commodity plastics like PP and PE). Two main types of CBAs are available: endothermic and exothermic agents. Endothermic agents like citric acid and sodium bicarbonate which adsorb energy during their decomposition. Some disadvantages like auto-cooling and post-curing are also necessary in order to complete the reaction. The main advantage of endothermic blowing agents are they have decomposition temperatures of 160 and 210°C, almost same as the processing temperatures of commodity plastics like PP and PE and also provide higher gas yields of 120 cc/g at STP. Exothermic blowing agents as the name indicates gases during their decomposition (mostly $N_2$, $CO_2$, carbon monoxide and ammonia) and also release considerable amounts of heat during their decomposition thus giving rise to formation of hot spots in the melt.

## 7.3 Super Critical Foaming

Generating foams using a physical blowing agent consists of saturating the polymer at a certain pressure and temperature followed by thorough mixing and equilibrium form the main steps in super critical fluid assisted foaming of polymers. This equilibrated mixture of the gas and polymer is subjected to a sudden thermodynamic change (either increase in temperature or a sudden and catastrophic drop in pressure) which causes the saturated gas to escape leaving behind a cellular structure. This process can result in both closed-cell or rigid foams if the cell membranes around the bubble remain intact or open-cell or flexible foams if the membranes rupture. The rate controlling steps in this process are:

- Degree of solubility of the gas in the polymer
- Degree of saturation achieved.
- Suddenness of the thermodynamic changes.

Of these the suddenness of the thermodynamic change is important. A schematic diagram showing the dependence of viscosity (which in fact is dependent on viscosity) is shown in the Fig. 7.3.

Since supercritical foaming of polymers are always performed under some sort of 'driven extensional flows' like extrusion or injection molding, the elongational or extensional deformation of the polymer melt also controls the foaming process.

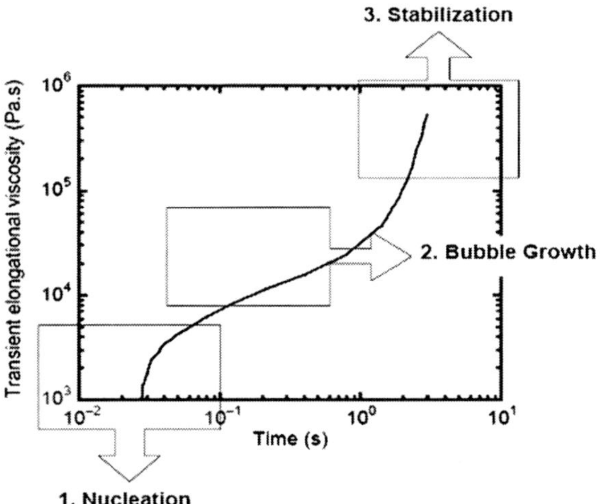


**Fig. 7.3** Dependence of various stages of microcellular foam formation with transient elongational viscosity in supercritical foaming of polymers

At present there are three choices of physical foaming agents: HFCs and HCFC (hydrogenated fluorocarbons and chlorofluorocarbons), alcohols and inert gases like $CO_2$ and $N_2$. Since HFCs and HCFCs are banned by Montreal protocol their usage will be banned by 2012 in many developed countries and by 2020 in developing countries. Alcohols are rarely soluble in commodity plastics and are poor foaming agents. So, the obvious choice is between $CO_2$ and $N_2$. As mentioned earlier $CO_2$ is superior since it's soluble in plastics is about four times higher than $N_2$. More ever, $CO_2$ also has advantages like being non-flammable, chemically stable and inert and economically low cost. However usage of $CO_2$ has challenges (but not disadvantages) like necessity of high-pressure operation, fast gas escape, dimensional instability during the foaming process.

Recently, foaming technology has penetrated into the research and development of wood-fiber plastic composite products [1–12]. As a result that their drawbacks such as higher density, lower ductility, and poor impact resistance compared to neat plastics and/or solid wood could be overcome with the presence of cellular structure within the composites. Foaming of plastic/wood-fiber composites can be produced by utilizing either a chemical or physical blowing agent. A pressure-quench method described by Goel and Beckman is widely used for making microcellular polymers via supercritical carbon dioxide ($scCO_2$) [13]. They found that the microcellular structure could be achieved by rapid depressurization to allow the cells nucleation and growth as in the batch process after saturating polymers with $scCO_2$.

Supercritical method of foaming is particularly advantageous in WPC since the presence of wood particles act as nucleating agents that act as 'hot spots' and reduce the nucleation energy. However it must be mentioned that the distribution of these wood fibers should be fine. Therefore recently additional fillers of layered structure like talc, kaolin, diatomaceous earth etc. have been used.

## 7.4 Other Techniques

Besides these two commonly used methods, some specialized and high precision methods (especially in biological applications like scaffolds in tissue engineering) of producing foams are:

- Phase separation.
- Precipitation with a compressed fluid anti-solvent.
- In situ polymerization with different monomers followed by subsequent extraction of one phase
- Salt extraction wherein a water soluble salt is blended into polymers and then washed out to form micro-porous open cell foams etc. have been used to achieve more uniform and controlled pore structure.

However, extrusion process with either super critical or with CBA is the most commonly used for production of microcellular composites on industrial scales. But extrusion process to create microcellular composites is highly empirical, complex and difficult to control process. Most companies guard their expertise very passionately. Many factors namely temperature, concentration of the blowing agent, mechanical stirring etc. also makes the difference between open cell and closed cell in extrusion foaming. Higher temperature gives rise to open structures since high foaming temperature prevents the freezing of cell structure before the rupture of cell walls. However, very high foaming temperature may cause cell collapse due to rapid loss of blowing agent (both in CBA and super critical fluid).

## 7.5 Microcellular Foaming Procedure

In general foaming process consists of three main steps:

- Preparation of plastic, wood fiber and blowing agent mixture
- Cell nucleation
- Cell growth.

There are basically two main categories of producing microcellular plastic foams by extrusion.

- Controlled foam extrusion process
- Free foaming extrusion.

In controlled foam extrusion process, the schematic diagram of which is shown below, the wood/polymer mixture is melted in an extruder followed by injection of blowing agent which mixes with the polymer matrix and dissolves in it. This wood polymer melt containing blowing agent is subsequently forwarded to the head of an extruder which is partially plunged with a torpedo or mandrel. This mixture is rapidly cooled in a water cooled shaper connected to the head of an extruder. The foaming process in this procedure proceeds from inwards towards the outer layer. Therefore the outer skin layer has less or no microcells (which is cooled more rapidly than the inner core). In this process both cell nucleation and cell growth occurs within the shaper. The above procedure is commercially available as "Celuka process" and is widely used in industry. However there are many disadvantages of this Celuka process.

- First is the formation of solid or less foam outer skin layer which needs to be cut thereby leading to wastage of raw materials. However this problem can be overcome by a optimal design of the die.
- Secondly since the foaming process starts from core to outwards, the cell density and cell distribution at core seems to be higher when compared to skin. This problem can be overcome by use of special type of screws. (A detailed discussion of screw characteristics is given in the later parts of this chapter) (Fig. 7.4).

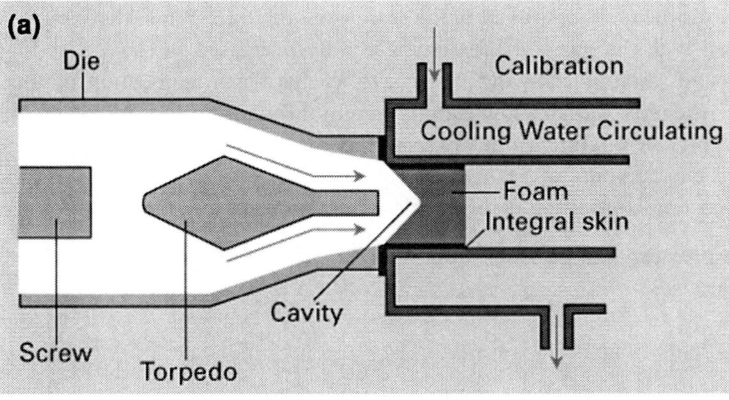


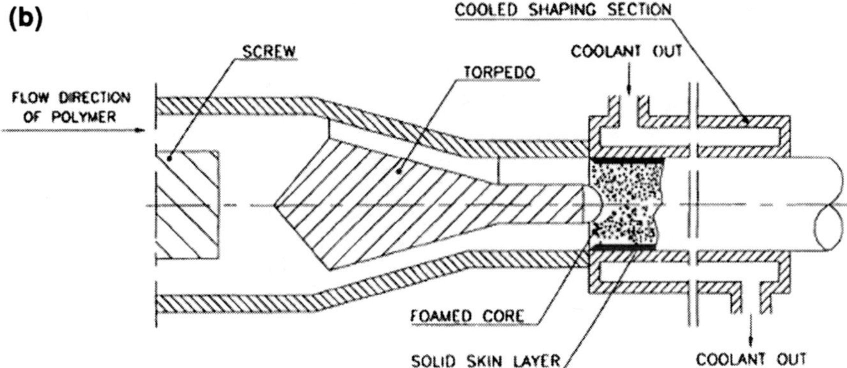


**Fig. 7.4** Schematic representation of Celuka process (**a**, **b**)

### 7.5.1 Free Foaming Extrusion Process

In contrast to controlled foam extrusion, free foaming extrusion as the name itself indicates the foaming of wood/polymer mixture is allowed to expand freely without any constraints. A schematic representation of this system is shown in Fig. 7.5.

The rate controlling step in this process is the temperature of the melt at the die exit. If the temperature of the melt is too high, there will be a sudden burst of bubbles within the polymer matrix, thereby leading to open cell structure. On the other hand, if the temperature is low enough there will not be sufficient foaming. So, some researchers tried gradual reduction of temperature but it substantially slows down the whole process thereby affecting productivity.

### 7.5.2 Batch Process

In batch process, the wood/polymer mixture is placed in a high pressure chamber connected to a gas reservoir of $N_2$ or $CO_2$. Under high pressures and sufficient times gas is adsorbed and dissolved in the composite mixture. When the sample is fully saturated with the gas, the pressure is rapidly decreased to cause a sudden drop in pressure, thereby allowing the gas to escape. This application of thermodynamic instability initiates nucleation, thereby forming billions of microcells in the wood/polymer mixture. In this process there is not much cell growth due to the fact that a large amount of gas escapes the system at very short periods of time (Fig. 7.6). The rate controlling steps in this whole procedure is the

- Saturation pressure and pressure drop
- Temperature
- Time.

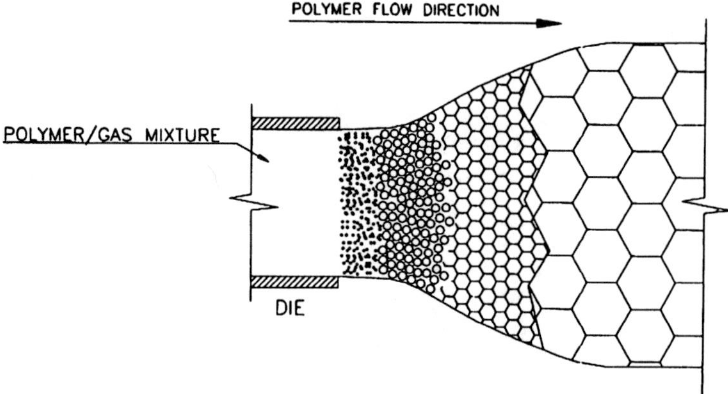


**Fig. 7.5** Free foaming extrusion process

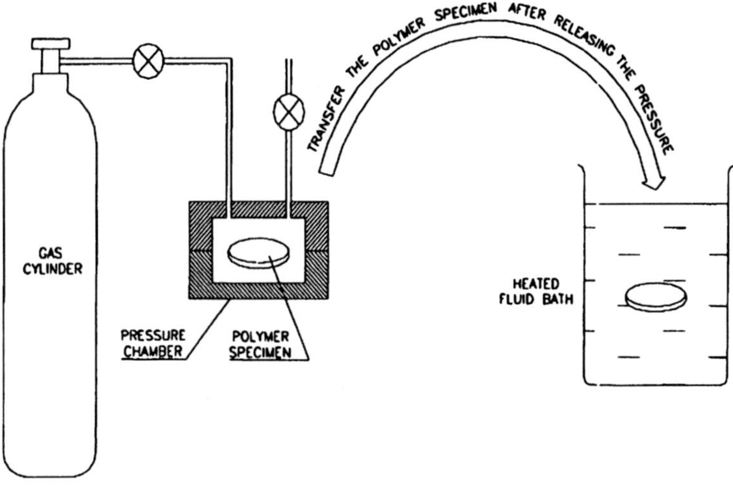


**Fig. 7.6** Conventional batch process of microcellular foam production

All the above variables can be easily controlled thereby giving the manufacturer great deal of flexibility in controlling the size and shape of the microcells. For example, optimum choice of temperature is generally chosen to be lowest possible to make the cell growth step highly controllable. However lower temperatures can also have disadvantage, when the nuclei are generated, their growth is retarded due to high stiffness of the polymer composite (in fact in batch processing, the chances of cell coagulation is minimal considering the difficulties in nucleation). Therefore by careful modulation of temperature and time cell growth can be controlled.

However the overall productivity by the batch process is adversely affected by the long times required for saturation of gas in the plastic. This is due to low rate of gas dissolution and diffusion in polymer systems. This problem is especially serious in wood/polymer composites, since the wood flour particles are relatively large, thereby leading to localized dissolution of gas. Therefore the need for evaluation of PVT (Pressure–Volume–Temperature) data is essential. Traditional measurement of PVT data by dilatometer is time consuming and requires lots of manpower. Herein we describe two simple and 'inline' methods to measure PVT data in WPC. These methods are based on extruders and the data obtained from these sets of experiments can be safely and successfully utilized for any sort of foaming techniques (batch or continuous).

#### 7.5.2.1 PVT Measurement Using a Gear Pump

The inherent characteristic of gear pump is positive displacement which characteristic can be used to measure the volumetric flow rate of the polymer melt. By correct and accurate measurement of volumetric ($Q$) and mass flow rate ($m$) during

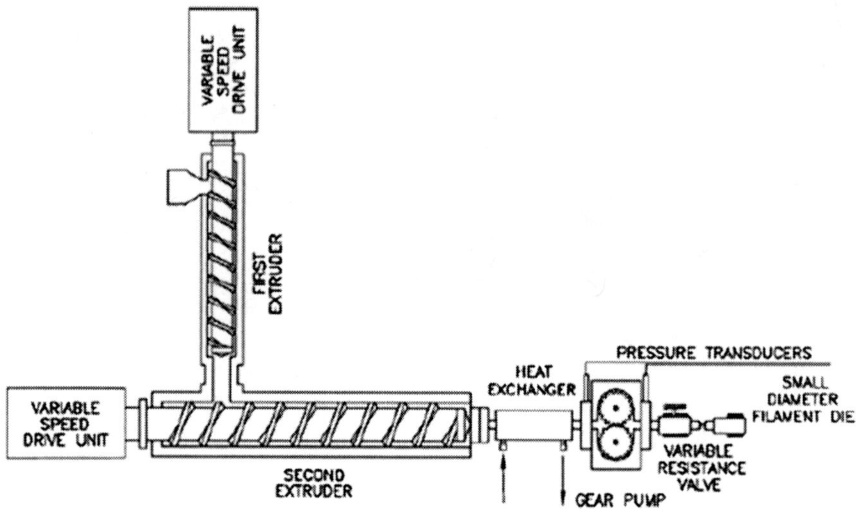


**Fig. 7.7** Schematic representation of PVT measurement using a gear pump

extrusion at a specific temperature and pressure, the specific volume can be calculated from the following equation:

$$\upsilon = \frac{1}{\rho} = \frac{Q}{m}. \tag{7.1}$$

The Fig. 7.7 shows the experimental setup for expanded WPC.

From the Fig. 7.7 it can be observed that two extruders are in tandem. The first extruder is a polymer melting extruder which will melt and plasticates the polymer pellets into a melt. The second extruder maintained at lower temperatures sequentially reduces the temperature while constantly building up the pressure. This buildup of pressure is achieved by special screw designs. The second extruder acts as temperature and pressure stabilizer. The next step is calibration of gear pump which can be achieved by standard techniques. The mass flow rate from both the extruders can be calculated by collecting the extruded melt for a fixed period of time and measuring the weight. Pressure can be easily controlled since we are using gear pumps wherein both the upstream and downstream pressures can be easily controlled.

The polymer pellets were first fed into the hopper of the tandem extruder. The extrudate from the first extruder passes through the second extruder and also through the heat exchanger thus facilitating an accurate control of the temperature of the melt. When a steady state condition was achieved at a particular temperature, pressure and gear speed, the mass flow rate was calculated. The specific volume can be calculated as the ratio of $Q$ and $m$, which were controlled by the rotational speed of the gear pump. Measured pressure versus specific volume was plotted as shown in the Fig. 7.8.

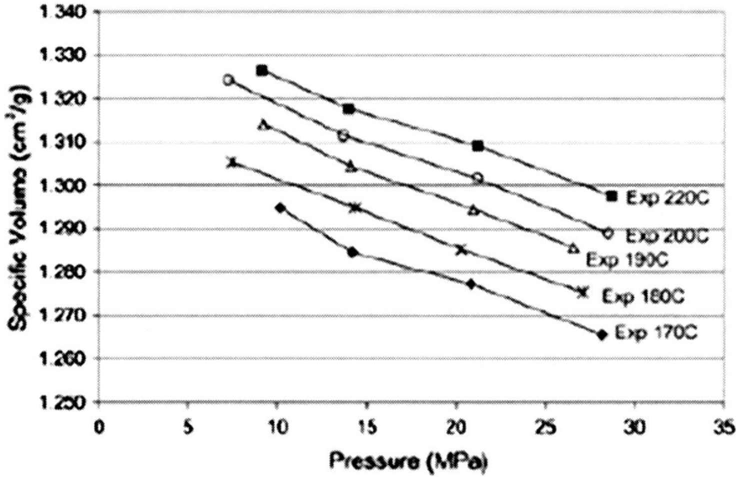


**Fig. 7.8** Measured PVT data for linear PP

The plot provides a quick and accurate measurement of PVT of polymer melt. The PVT data of WPC can also be calculated by varying the concentration of wood flour, the amount of dissolved gas etc. in the polymer composite thereby providing the polymer engineer an accurate measure of solubility of the foaming gases in the polymer melts.

However certain corrections need to be used in order to decouple the cell geometry in foams. The modified cell model (MCM) is described as:

$$\frac{P\bar{\upsilon}}{\bar{T}} = \frac{\bar{\upsilon}^{1/3}}{\bar{\upsilon}^{1/3} - 0.8909q} - \frac{2}{\bar{T}}\left(\frac{1.2045}{\bar{\upsilon}^2} - \frac{1.011}{\bar{\upsilon}^4}\right) \tag{7.2}$$

$$\text{where, } \bar{\upsilon} = \frac{\upsilon}{\upsilon^*} \text{ and } \upsilon^* = \sigma^3 \tag{7.3}$$

$$\bar{T} = \frac{T}{T^*} \text{ and } T^* = \frac{s\eta}{ck} \tag{7.4}$$

$$\bar{P} = \frac{P}{P^*} \text{ and } P^* = \frac{ckT^*}{\upsilon^*}. \tag{7.5}$$

#### 7.5.2.2 PVT Measurement Using a Capillary Rheometer

A capillary die is mounted in the downstream of a molten wood polymer composite thereafter the volumetric flow rate and pressure drop is measured. For WPC within the capillary die, the modified power law model as suggested by the following expression shown below has been used

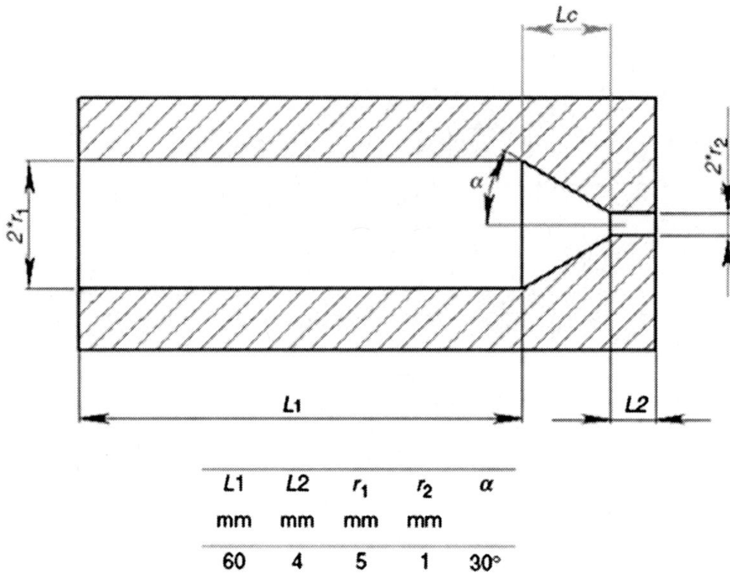


| L1 | L2 | $r_1$ | $r_2$ | α |
|---|---|---|---|---|
| mm | mm | mm | mm | |
| 60 | 4 | 5 | 1 | 30° |

**Fig. 7.9** Cross section view of capillary die

$$\tau = \tau_Y + K.\dot{\gamma}^n \text{ for } |\tau| \geq |\tau_Y| \tag{7.6}$$

$$\dot{\gamma} = 0 \text{ for } |\tau| < |\tau_Y|. \tag{7.7}$$

Typically the capillary die consists of two capillary sections and a conical outlet section as indicated Fig. 7.9.

The resulting pressure drop in the capillary die can be the sum of pressure drop in the three sections:

$$\Delta P = \Delta P_{\text{cap1}} + \Delta P_{\text{cap2}} + \Delta P_{\text{cap3}}. \tag{7.8}$$

The pressure drop is various sections are calculated by the following expressions:

$$\Delta P_{\text{capillary}} = \frac{2LK}{r^{3n-2}\left[n\pi(3n+1)^{1/n}\right]^{n-1}}Q \tag{7.9}$$

$$\Delta P_{\text{conic}} = \frac{2K}{3n\tan\alpha}\left(\frac{3n+1}{n\pi r_2^3}\right)^n\left[1-\left(\frac{r_2}{r_1}\right)^{3n}\right]Q^n \tag{7.10}$$

where $Q$ is the volumetric flow rate. So, the total pressure drop occurring within the die in all sections is

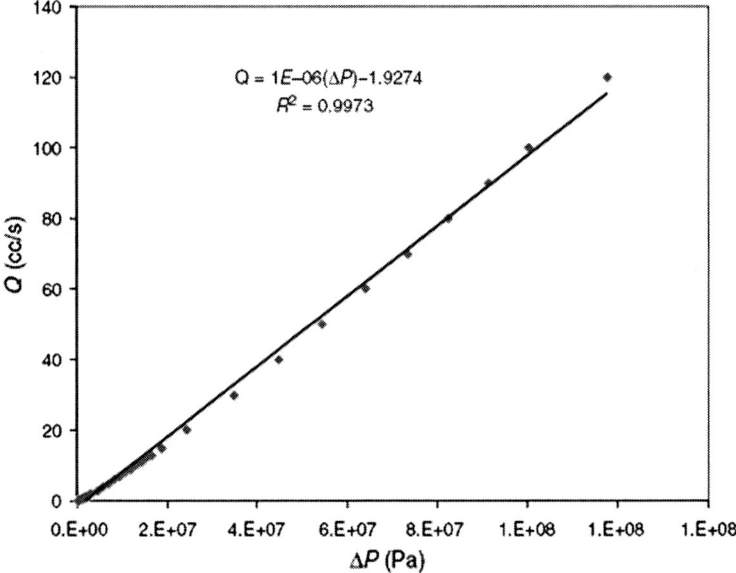


**Fig. 7.10** Linear relationship between pressure drop and volumetric flow rate in 40% WPC based on HDPE

$$\Delta P = ABQ + CQ^n \tag{7.11}$$

where $A$, $B$ and $C$ are coefficients related to the geometry of the die and are calculated by the following expressions:

$$A = \frac{2K}{\left(n\pi(3n+1)^{1/n}\right)^{n-1}} \tag{7.12}$$

$$B = \frac{L_1}{r_1^{3n-2}} + \frac{L_2}{r_2^{3n-2}} \tag{7.13}$$

$$C = \frac{2K}{3n\tan\alpha}\left(\frac{3n+1}{n\pi r_2^3}\right)^n \left[1 - \left(\frac{r_2}{r_1}\right)^{3n}\right]. \tag{7.14}$$

For a typical HDPE based WPC the following linear relationship has been observed (Fig. 7.10).

By using any of the above methods one can obtain accurate PVT data which can be used to calculate the solubility of the foaming gas in order to better understand the foaming technique.

### 7.5.3 Factors and Problems Affecting Foaming of Wood Polymer Composites

Essentially microcellular foaming of WPC should have the following essential processing mechanisms:

- A mechanism for complete dissolution of large amounts of blowing gas into the polymer. This problem is very important in case of WPCs since, normally 50/50 blends of wood and polymer are used and the gas dissolves only in the polymer phase and the wood phase is unaffected which restricts the overall dissolution of gases in the composite. Therefore very high pressures and large soaking times are essentially used. However utilization of higher pressures may cause defragmentation in the wood-polymer interfaces thereby reducing performance characteristics like tensile strength and modulus. This defragmentation is a serious problem in case of thin WPC foams making the surface of the WPC foam very rough. However another set of problem arises if the interactions between the wood and polymer are too high: heterogeneous nucleation (discussed below).
- Wood powders are polar in nature whereas commodity plastics are non-polar. This results in non-uniform and incomplete wetting of wood fibers within the polymer matrix. These channels provide channels for fast escape of the gases. Consequently the apparent effective diffusion of the gas is enhanced.
- Cell nucleation and bubble growth are the fundamental rate controlling steps in any foaming process. During foaming of wood fiber reinforced polymer composites, because of high loadings of wood fibers, the potential of for heterogeneous nucleation occurs at the solid melt interface due to increase in free energy of the system caused by reduction in the surface tension at the solid fiber-viscoelastic polymer interface. If the interface between the fiber and polymer is weak the gas is entrapped in the micro-voids at the interfaces giving rise to a new form of heterogeneous nucleation, termed as 'self nucleation'. Another kind of self nucleation occurs if high temperatures are used during processing. At temperatures of 100–110°C wood fibers release moisture (this problem is acute in humid environments wherein the wood particles quickly adsorb the atmospheric moisture). This water vapor has low dissolution in polymer (due to non-polar nature of the polymer matrix) thereby entrapping the water vapor in the interfacial regions of the wood-polymer interfaces. So the usage of 'bone dry' wood fibers and wood powders is recommended for WPC foam production. Yet another type of 'self nucleation' also arises. Since commodity plastics are processed at temperatures of 160–190°C, the wood fibers release some volatile gases. (A seasoned extruder operator will definitely know this). These vapors however pleasant they smell do not dissolve in plastics and if are not allowed to escape quickly get trapped in the interfacial region.
- If one uses chemical blowing agents some of the CBA gets adhered to the surface of the dispersed wood fibers and when the composite is foamed there

will be localized blow out of wood fibers causing surface instabilities. This problem will be more acute in case of wood flour/polymer when compared to wood fiber/polymer foams.

- The usage of chemical blowing agents especially exothermic kinds like ADC and AZDC have been known to provide many advantages when compared to physical blowing agents like super critical gases. CFA usage to foam WPC has gained interest because properties such as insulation values, shrinkage and distortion and stiffness can be controlled effectively since CFAs can provide consistent process control and nucleating effects which can solve the moisture problems and vastly improve mechanical properties.

# References


1. Zhang, S., Rodrigue, D.: Preparation and morphology of polypropylene/wood flour composite foams via extrusion. Polym. Compos. **26**, 731–738 (2006)
2. Li, Q., Matuana, L.M.: Foam extrusion of high density polyethylene/wood-flour composites using chemical foaming agents. J. Appl. Polym. Sci. **88**, 3139–3160 (2002)
3. Matuana, L.M., Mengeloglu, F.: Microcellular foaming of impact-modified rigid PVC/wood-flour composites. J. Vinyl Addit. Technol. **7**(2), 67–76 (2001)
4. Matuana, L.M., Park, C.B., Balatinecz, J.J.: Processing and cell morphology relationships for microcellular foamed PVC/wood-fiber composites. Polym. Eng. Sci. **37**(7), 1137–1147 (1997)
5. Rachtanapun P., Selke S.E., Matuana L.M.: Microcellular foam of polymer blends of HDPE/PP and their composites with wood fiber. J. Appl. Polym. Sci. **88**(12), 2842–2860 (2003)
6. Bledzki, A.K., Faruk, O.: Injection moulded microcellular wood fibre-polypropylene composites. Compos. Part A. **37**, 1368–1367 (2006)
7. Bledzki, A.K., Faruk, O.: Influence of processing temperature on microcellular injection-moulded wood-polypropylene composites. Macromol. Mater. Eng. **29**(10), 1226–1232 (2006)
8. Gosselin, R., Rodrigue, D., Riedl, B.: Injection molding of postconsumer wood–plastic composites I: morphology. J. Thermoplast. Compos. Mater. **19**(6), 639–667 (2006)
9. Gosselin, R., Rodrigue, D., Riedl, B.: Injection molding of postconsumer wood–plastic composites II: mechanical properties. J. Thermoplast. Compos. Mater. **19**(6), 659–669 (2006)
10. Mengeloglu, F., Matuana, L.M.: Manufacture of rigid PVC/wood-flour composite foams using moisture contained in wood as foaming agent. J. Vinyl Addit. Technol. **8**, 264–270 (2002)
11. Zhang, H., Rizvi, G.M., Park, C.B.: Development of an extrusion system for producing fine-celled HDPE/wood–fiber composite foams using $CO_2$ as a blowing agent. Adv. Polym. Technol. **23**(4), 263–276 (2004)
12. Rizvi, G.M., Park C.B., Matuana, L.M.: Foaming of PS/wood fiber composites using moisture as a blowing agent. Polym. Eng. Sci. **40**(10), 2124–2132 (2000)
13. Goel, S.K., Beckman, E.J.: Generation of microcellular polymeric foams using supercritical carbon dioxide. II Cell growth and skin formation. Polym. Eng. Sci. **34**, 1148–1166 (1994)

# Chapter 8
# Wood Plastic Composite Foam Applications

## 8.1 Introduction

Wood fiber/plastic composites (WPCs) utilize fibers as reinforcing filler in the polymer matrix and are known to be advantageous over the neat polymers in terms of the materials cost and mechanical properties such as stiffness and strength. Wood fiber reinforced polymer composites are microcellular processed to create a new class of materials with unique properties. Most manufacturers are evaluating new alternatives of foamed composites that are lighter and more like wood. Foamed wood composites accept screws and nails like wood, more so than their non-foamed counterparts. They have other advantages such as better surface definition and sharper contours and corners than non-foamed profiles, which are created by the internal pressure of foaming. The microcellular wood fiber reinforced polymer composites can be obtained by different processes (batch, injection molding, extrusion, and compression molding process) by using physical or chemical foaming agent.

Matuana et al. [1–5] investigated the processing of microcellular-foamed structures in PVC/wood fiber (silane treated) composites by a batch-foaming process. They have established the relationships between cell morphology and processing conditions, as well as between the cell morphology and mechanical properties. It is seen that cell densities show a decreasing tendency with the increase of foaming temperatures. Matuana et al. [6] also investigated microcellular foam of polymer blends of HDPE/PP with wood fiber in a batch process. HDPE/wood composites had a reasonably high void fraction at high foaming times compared to PP/wood composites and HDPE/PP blend-wood composites. The batch-foaming process used to generate cellular foamed structures in the composites is not likely to be implemented in the industrial production of foams because it is not cost-effective. The microcellular batch-foaming process is time-consuming because of the multiple steps in the production of foamed samples [6]. In order to overcome the shortcomings of the batch process, a cost-effective, continuous microcellular

J. K. Kim and K. Pal, *Recent Advances in the Processing of Wood–Plastic Composites*, Engineering Materials, 32, DOI: 10.1007/978-3-642-14877-4_8, 

process (injection molding, extrusion, and compression molding process) was developed based on the same concept of thermodynamic instability that is found in the batch process.

Bledzki et al. [7] investigated the influence of endothermic (sodium bicarbonate, citric acid), exothermic (mostly azodicarbonamide) and endo/exothermic chemical foaming agent on microcellular wood fiber reinforced polypropylene (PP) composites in injection moulding process, it was seen that an exothermic foaming agent shows finer microcellular structures and better mechanical properties compared to endothermic and endo/exothermic chemical foaming agents. Bledzki et al. [8] reported that the melt flow index of PP and a variation of injection parameters (mold temperature, front flow speed, and filling quantity) have a great influence on the properties and structure of the wood fiber/PP composite foams. Bledzki et al. [9] also demonstrated that the wood fiber type and length strongly affect the microcellular structures. Finer wood fibers are correlated with a finer microcellular structure. Microcellular soft wood fiber-PP composites were also prepared by Bledzki et al. [10] in a box part by an injection molding process, and a comparative study of cell morphology, weight reduction, and mechanical properties between the box part and a panel shape using soft wood fiber/PP composite foams by considering different processing temperatures. The cell morphology of the injection molded box part differed in parts, with the area near the injection point showing a finer cellular structure than that of the areas far from the injection point area. As a result, the mechanical properties also differed in parts of the areas. Gosselin et al. [11, 12] reported the relations between the processing conditions with morphology and mechanical properties of injection molding of postconsumer wood–plastic composites.

Extrusion processes continuously devolatize wood fibers and other natural cellulosic materials and mix with plastics. A suitable combination of process variables was necessary for limiting the thermal degradation of the wood and natural fibers. Rigid PVC-wood fiber composites foamed in a continuous extrusion process were investigated by Matuana et al. [13–15] the effects of wood fiber moisture content, all-acrylic foam modifier content, CFA content, and extruder die temperature on the foamed composites structure and properties were studied. Zhang et al. [16] have experimented with tow system configurations (tandem extrusion system vs. single extruder system) for wood fiber-polymer composites to demonstrate the system effect on the cell morphology and foam properties. Polystyrene (PS)-wood fiber foamed composites were investigated using moisture as a foaming agent [17]. HDPE-wood fiber foamed composites were also investigated by considering the effect of CFA (endothermic and exothermic) and the influence of critical processing temperature on the cell morphology [18, 19]. Matuana et al. [20] examined the extrusion foaming of PP-wood fiber composites using a factorial design approach to evaluate the statistical effects of materials used and processing conditions on the void fraction. Nowadays, nanoparticles (i.e., clay) are used in microcellular wood fiber reinforced polymer composites. Guo et al. [21] showed the addition of clay reduces the cell size, increases the cell density of metallocene polyethylene/wood fiber composites in the extrusion foaming using a CBA. Lee et al. [22] studied the effects of nanoclay on the extrusion foming of polyethylene/wood fiber nanocomposites using $N_2$ as blowing agent. Rodrigue et al. [23] investigated the effect of wood powder on the polymer foam nucleation of wood-low density PE composites in an extrusion process and reported that the wood particles act as nucleating agents to substantially reduce cell size and increase cell density.

## 8.2 Wood–fiber/Plastic Composite Foams (WPCs Foams)

Wood fiber has recently received considerable attention as a filler to reinforce plastics. This trend is driven by the continuous increase of oil price and a strong desire to use more environmental methods of plastic production. Compared to inorganic fillers wood fiber offers the advantages of low density and biodegradability [24–26]. However, the density of wood fiber is still higher than that of most thermoplastics [3], and a manufacturing process that can lower the density (and weight) of wood-fiber/plastics (WPC) composite parts while maintaining its functional properties is definitely desirable to end users. Foaming technology presents such a possibility [3, 8, 14, 16]. In this part, firstly, the expanded WPC composite pellets were produced by batch-foaming process, and then WPC foamed board was made by compression molding by some researchers.

## 8.3 Preparation Wood–fiber/PP Composites by Twin Screw Extruder

Wood–fiber/PP composites were prepared with PP 65 phr, PP-g-MA 5 phr, Wood–fiber 30 phr, and SEBS-g-MA 5 phr, respectively, which extruded by co-rotating intermeshing twin-screw extruder (Bau-Tech, Korea). It has a screw diameter of 19 mm and the distance between screw axes is 18.4 mm with L/D ratio of 40. The screw speed was fixed 150 rpm while the cylinder temperature was maintained at 150, 165, 175, and 180°C from the hopper to the die. The extrudate was pelletized and dried under vacuum at 80°C for 24 h to remove any residual water.

## 8.4 Preparation of Expanded Wood–fiber/PP Composite Pellet

Expanded wood–fiber/PP composite pellet was prepared in a batch process. A schematic of the batch-foaming process is shown in Fig. 8.1. The extruded wood–fiber/PP composite pellet was enclosed high-pressure vessel. The vessel was flushed with low-pressure $CO_2$ for about 3 min and pressurized to the saturated

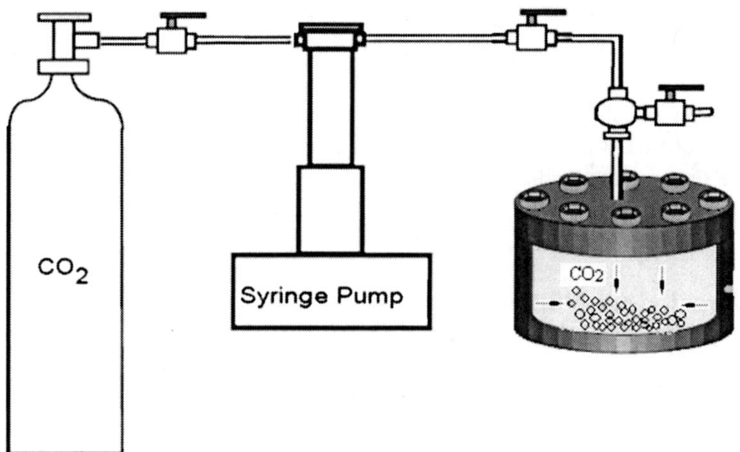


**Fig. 8.1** The schematics of batch-foaming process

vapor pressure $CO_2$ at room temperature and preheated to desired temperature. Afterward, the pressure was increased to the desired pressure by a syringe pump and maintained at this pressure for 1 h to ensure equilibrium absorption of $CO_2$ by the samples. After saturation, the pressure was quenched atmospheric pressure within 3 s and the samples were taken out. Then foam structure was allowed to full growth during rapid depressurization.

## 8.5 Preparation of WPC Foamed Board

The expanded wood–fiber/PP composite pellets were placed in a 170 × 150 × 3.5 mm mold together and loaded in a hydraulic hot press for 10 min at pressure 10 MPa and different temperature. After the conditions were reached, the pressure was released and the mold was taken out of the hot press immediately. Then we unloaded the mold, quenched it to room temperature and removed the WPC foamed board from the mold, as shown in Fig. 8.2.

## 8.6 Effect of Temperature on the Density of WPC Foamed Board

In order to study the effect of saturation temperature on the density of foamed board, the saturation pressure was fixed at 16 MPa, the Wood–fiber/PP composite pellet was foamed at different saturation temperature, and then the foamed board was prepared by compression molding. The upper and lower plate temperature of hot press was setting at 170 and 150°C, respectively. From Fig. 8.3, it can be

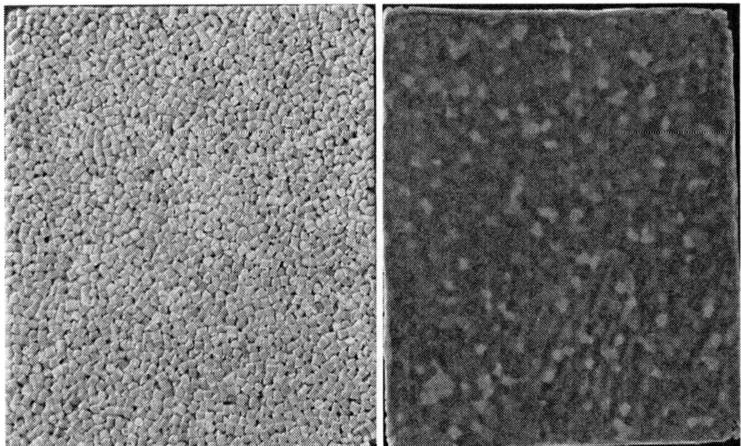

**Fig. 8.2** Foamed board made from WPC

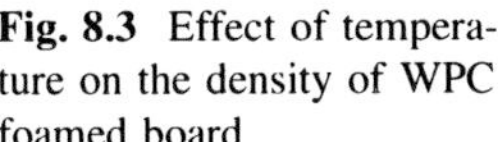

**Fig. 8.3** Effect of temperature on the density of WPC foamed board

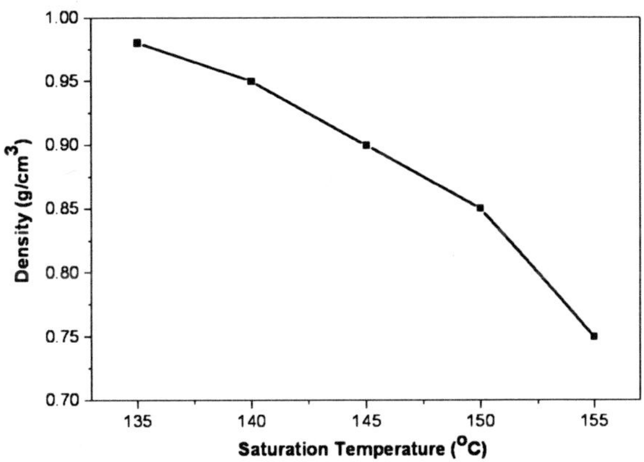


observed that the density of board decrease with increasing of saturation temperature. As aforementioned in Chap. 7, as temperature increases, the viscosity of the polymer matrix decreases facilitating the cells grow up, so the density of expanded WPC pellet decreased, and the corresponding density of board decreased.

## 8.7 Effect of Pressure on the Density of WPC Foamed Board

In order to study the effect of saturation pressure on the density of foamed board, the saturation temperature was fixed at 150°C, the wood–fiber/PP composite pellet was foamed at different saturation pressure, and then the foamed board was prepared by

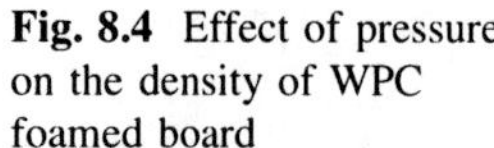

**Fig. 8.4** Effect of pressure on the density of WPC foamed board

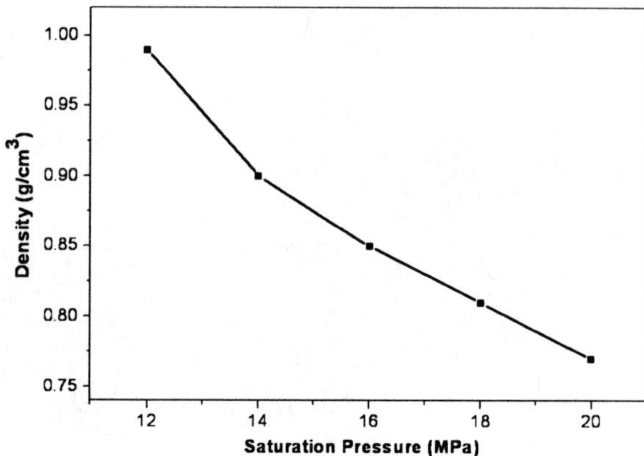


**Fig. 8.5** Effect of plate temperature on the density of WPC foamed board

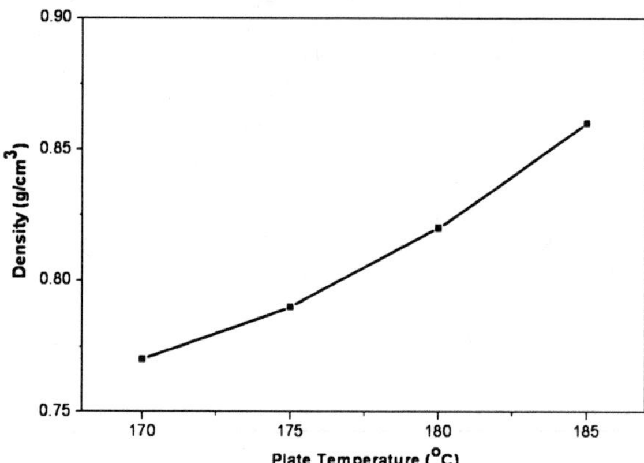


compression molding. The upper and lower plate temperature of hot press was setting at 170 and 150°C, respectively, from Fig. 8.4, it can be observed that the density of board decrease with increasing of saturation pressure. As aforementioned in last chapter, as pressure increases, the amount of $CO_2$ dissolved in samples increases. As a result, the melt strength of the matrix is weaker, so the sample is relatively soft and its deformability is large, so the density of expanded wood–fiber/PP composite pellets decreased, and the corresponding density of board decreased.

## 8.8 Effect of Plate Temperature on the Density of WPC Foamed Board

In order to study the plate temperature on the density of foamed board, the expanded WPC pellets were obtained at a given saturation temperature and pressure of 150°C and 20 MPa, respectively, and then the foamed board was prepared by compression molding at different upper plate temperature of hot press. From Fig. 8.5, it can be seen that, the density of foamed board increased with increasing of upper plate temperature. At higher temperature, there are more pellets will change from foam to solid, the density increased.

## References

1. Matuana, L.M., Mengeloglu, F.: Microcellular foaming of impact-modified rigid PVC/wood-flour composites. J. Vinyl Addit. Technol. **7**(2), 67–75 (2001)
2. Matuana, L.M., Park, C.B., Balatinecz, J. J.: Processing and cell morphology relationships for microcellular foamed PVC/wood–fiber composites. Polym. Eng. Sci. **37**(7), 1137–1147 (1997)
3. Matuana, L.M., Park, C.B., Balatinecz, J. J.: Effect of cell morphology on the properties of microcellular foamed PVC/wood–fiber composites. Cell. Microcell. Mater. **76**(5), 1–16 (1996)
4. Matuana, L.M., Park, C. B., Balatinecz, J. J.: Cell morphology and property relationships of microcellular foamed PVC/wood–fiber composites. Polym. Eng. Sci. **38**(11), 1862–1872 (1998)
5. Matuana, L.M., Park, C. B., Balatinecz, J. J.: Characterization of microcellular foamed PVC/cellulosic-fibre composites. J. Cell. Plast. **32**(9), 449–469 (1996)
6. Rachtanapun, P., Selke, S. E., Matuana L. M.: Microcellular foam of polymer blends of HDPE/PP and their composites with wood fiber. J. Appl. Polym. Sci. **88**(12), 2842–2850 (2003)
7. Bledzki, A.K., Faruk, O.: Injection moulded microcellular wood fibre-polypropylene composites. Compos. A **37**, 1358–1367 (2006)
8. Bledzki, A.K., Faruk, O.: Effects of the chemical foaming agents, injection parameters, and melt-flow index on the microstructure and mechanical properties of microcellular injection-molded wood–fiber/polypropylene composites. J. Appl. Polym. Sci. **97**, 1090–1096 (2005)
9. Bledzki, A.K., Faruk, O.: Microcellular injection molded wood fiber–PP composites: Part II—effect of wood fiber length and content on cell morphology and physico-mechanical properties. J. Cell. Plast. **42**, 77–88 (2006)
10. Bledzki, A. K., Faruk, O.: Influence of processing temperature on microcellular injection-moulded wood-polypropylene composites. Macromol. Mater. Eng. **29**(10), 1226–1232 (2006)
11. Gosselin, R., Rodrigue, D., Riedl, B.: Injection molding of postconsumer wood–plastic composites I: morphology. J. Thermoplast. Compos. Mater. **19**(6), 639–657 (2006)
12. Gosselin, R., Rodrigue, D., Riedl, B.: Injection molding of postconsumer wood–plastic composites II: mechanical properties. J. Thermoplast. Compos. Mater. **19**(6), 659–669 (2006)
13. Mengeloglu, F., Matuana, L.M.: Manufacture of rigid PVC/wood-flour composite foams using moisture contained in wood as foaming agent. J. Vinyl Addit. Technol. **8**, 264–270 (2002)
14. Mengeloglu, F., Matuana, L.M.: Foaming of rigid PVC/wood-flour composites through a continuous extrusion process. J. Vinyl Addit. Technol. **7**, 142–148 (2001)
15. Mengeloglu, F., Matuana, L.M.: Mechanical properties of extrusion-foamed rigid PVC/wood-flour composite. J. Vinyl Addit. Technol. **9**, 26–31 (2003)
16. Zhang, H., Rizvi, G. M. and Park, C. B.: Development of an extrusion system for producing fine-celled HDPE/wood–fiber composite foams using $CO_2$ as a blowing agent. Adv. Polym. Technol. **23**(4), 263–276 (2004)
17. Rizvi, G. M., Park C.B., Matuana, L.M.: Foaming of PS/wood fiber composites using moisture as a blowing agent. Polym. Eng. Sci. **40**(10), 2124–2132 (2000)

18. Matuana, L.M., Li, Q.: Foam extrusion of high density polyethylene/wood-flour composites using chemical foaming agents. J. Appl. Polym. Sci. **88**, 3139–3150 (2003)
19. Guo, G., Wang, K.H., Park, C.B., Kim, Y.S., Li, G.: Critical processing temperature in the manufacture of fine-celled plastic/wood–fiber composite foams. J. Appl. Polym. Sci. **91**, 621–629 (2004)
20. Matuana, L.M., Li, Q.: A factorial design applied to the extrusion foaming of polypropylene/wood-flour composites. Cell. Polym. **20**(2), 115–130 (2001)
21. Guo, G., Wang, K.H., Park, C.B., Kim, Y.S., Li, G.: Effects of nanoparticles on the density reduction and cell morphology of extruded metallocene polyethylene/wood fiber nanocomposites. J. Appl. Polym. Sci. **104**, 1058–1063 (2007)
22. Lee, Y.H., Kuboki, T., Park, C.B., Sain, M.: The effects of nanoclay on the extrusion foaming of wood fiber/polyethylene nanocomposites, Simha symposium on polymer physics and polymer nanocomposites 2007 symposium, Boucherville, QC, Canada, October 17–19.[No. 180] (2007)
23. Rodrigue, D., Souici, S., Twite-Kabamba, E.: Effect of wood powder on polymer foam nucleation. J. Vinyl Addit. Technol. **12**, 19–24 (2006)
24. Bledzki, A.K., Faruk O., Huque, M.: Physico-mechanical studies of wood fiber reinforced composites. Polym. Plast. Technol. Eng. **41**(3), 435–451 (2002)
25. Bledzki, A.K., Letman, M., Viksne, A., Rence, L.: A comparison of compounding processes and wood type for wood fibre-PP composites. Compos. A **36**(6), 789–797 (2005)
26. Matuana, L.M., Heiden, P.A.: Wood Composites. In: Kroschwitz J.I. (Ed.) Encyclopedia of Polymer Science and Technology. Wiley, New York (2004)

# Chapter 9
# Conclusions

Wood, plastic and the uses of these composites in our daily life has been described in first chapter followed by the surface treatment and additives used for preparation of wood–plastic composites in subsequent chapter.

In the third chapter, machinery and testing techniques for making of wood–plastic composites has been discussed. Because, the authors think that proper machinery is the main important part to produce a good quality WPC. Fourth chapter is described about the research work done by other researchers in recent past.

One of the objectives of this research is to achieving the effects of screw configuration, screw speed, silica content and various compatibilizer on the physico-mechanical and foaming properties of wood–fiber/PP composites. Wood–fiber/PP composites were produced on the intermeshing co-rotating twin screw extruder. Microcellular closed cell wood–fiber/PP composite foams were prepared using pressure-quench batch process method. First, an attempt has been made to determine the optimum conditions of extrusion that involve are screw configurations, screw speed. The experimental results showed that wood–fiber/PP composite prepared under the configuration C (Fig. 5.1) at screw speed of 150 rpm have higher mechanical properties and narrower cell size distribution, due to the improved dispersion of the wood–fiber in the composites. And then, the effect of the silica content on properties of wood–fiber/PP/silica composites showed that the relationship between silica content and mechanical properties was obvious. The property enhancements were controlled mainly by the extension of silica agglomeration. The foam density is governed by the combined effect of cell nucleation, cell growth, and cell coalescence, as a nucleation agent, the silica in wood–fiber/PP composite created a large amount of heterogeneous nucleation sites during foaming. The heterogeneous nucleations increased the cell density and decreased the cell size and foam density. This phenomenon disappears at high content of silica due to big agglomerates. Finally, the results of physico-mechanical properties of wood–fiber/PP composite with various compatibilizer showed that the PP-g-MA and SEBS-g-MA enhanced adhesion between the WF and PP matrix,

J. K. Kim and K. Pal, *Recent Advances in the Processing of Wood–Plastic Composites*, Engineering Materials, 32, DOI: 10.1007/978-3-642-14877-4_9,

Wood–fiber/PP composite with addition of PP-g-MA as compatibilizer showed highest tensile strength and stiffness. Composites with SEBS-g-MA showed the higher impact strength, elongation at break and tensile strength compared to composite systems with SEBS. The batch foaming result of PP/WF composites with various compatibilizer showed relationships between cell morphology and their crystallinity and stiffness. Higher crystallinity, showed higher stiffness and higher relative density.

The fifth part is concerned the improvement of flammability of wood–fiber/PP composites by use of ammonium polyphosphate (APP) and silica as flame retardants, meanwhile, the effects of APP and silica on the mechanical and foaming properties of the wood–fiber/PP composites were studied. Marginal reduction in the mechanical properties of the composites was found with addition of flame retardants, except for the tensile strength of small amount of silica filled wood–fiber/PP composite. APP showed effective flame retardancy for wood–fiber/PP composites based on LOI and CONE data. And APP decreased initial temperature of thermal degradation and promoted char formation of the composite. Silica has been shown to have a flame retardant synergistic effect with APP in wood–fiber/PP composite. The TGA data showed that addition silica can lead to the formation of a more thermally stable barrier, and the residual char of the system with silica is higher than that of without silica. Microcellular wood–fiber/PP composite foams were successfully prepared with batch foaming method using supercritical carbon dioxide as blowing agent. In microcellular composites, the cell size and relative density was a strong function of APP content. With increasing of APP content, the relative density decreased and then increased, the cell size increased. The addition of silica leads to smaller cell size and higher relative density. As the saturation temperature and pressure increased, the cell size increased and the relative density decreased.

In the last potential application, wood–fiber/PP composite foamed board was successfully produced. The effect of three processing parameters on the density of foamed board was investigated. Bothe increasing the saturation temperature and pressure led to lower density, whereas the density increased as the upper plate temperature of hot press increased. In conclusion, the density of wood–fiber/PP composite foamed board can be controlled by changing the processing parameters.

# Index

CPSIA information can be obtained at www.ICGtesting.com
Printed in the USA
LVOW072121120613

338347LV00005B/29/P